Volume II

Applied Physics
and Engineering

An International Series

Electric Probes in Stationary and Flowing Plasmas:
Theory and Application

Electric Probes in Stationary and Flowing Plasmas: Theory and Application

Paul M. Chung
Lawrence Talbot
Kenell J. Touryan

Springer-Verlag Berlin · Heidelberg · New York
1975

Paul M. Chung
University of Chicago

Lawrence Talbot
University of California, Berkeley

Kenell J. Touryan
Sandia Laboratories, Albuquerque, New Mexico

Library of Congress Cataloging in Publication Data

Chung, Paul Myung-Ha
Electric probes in stationary and flowing plasmas.

1. Plasma diagnostics. 2. Probes (Electronic instruments) I. Talbot, Lawrence, 1925–, joint author. II. Touryan, K. J., joint author. III. Title.
QC718.5.D5C48 530.4′4 74-8844

Softcover reprint of the hardcover 1st edition 1975

ISBN-13: 978-3-642-65888-4 e-ISBN-13: 978-3-642-65886-0
DOI: 10.1007/978-3-642-65886-0

Preface

The electric probe has long been used as a fundamental diagnostic tool for measuring the local properties of a plasma. Since Langmuir first developed the electric-probe technique in 1924, probes have been used to measure electron densities and temperatures in a wide variety of gaseous ionized media, such as electric discharges, afterglows, ionizing shock waves, flames, MHD, and plasma-jet flows, reentry vehicle flow fields, and atmospheric and space plasmas.

The first systematic account of modern theories of electric-probe behavior was given by Chen (1965), who also provided practical information on experimental techniques. A subsequent survey by Swift and Schwar (1970), which was representative of results contained in the literature through 1969, included additional information on some of the modern theories and on practical details of probe utilization.

The purpose of this volume is to supplement the previously mentioned two works by providing an account of a large body of the up-to-date information available on electric probes, particularly in the areas of transitional and continuum-flow phenomena, and by offering, for all domains of probe application, a critical appraisal of the more significant probe theories and experimental investigations in the literature.

The volume is intended for the engineer or scientist who desires a comprehensive survey of probe theories and who wishes to apply the available information to specific diagnostic situations encountered in the laboratory or in flight environments. The emphasis therefore is on practicality and usefulness rather than mathematical rigor. Due consideration is given to the particular limitations inherent in each analysis or experimental correlation described.

Chapter I begins with a discussion of the fundamental definitions and key parameters that govern the electric-probe measurement technique. Chapter II is devoted to collisionless and transitional probe behavior in quiescent as well as flowing plasmas. Chapter III covers the continuum regime of probe operation. Chapter IV deals with special topics such as probe surface integrity, and probe measurements in turbulent flows and magnetic fields. The equations that provide the theoretical basis for the analysis of probe behavior are developed in some detail in the Appendix.

The authors express their deep appreciation to their colleagues I. M. Cohen and S. H. Lam for reviewing critically all or parts of the draft of the manuscript and for many valuable suggestions. A very special note of thanks is due to Myrna L. Walla for patiently organizing and typing the seemingly endless versions of the original manuscript.

This work was supported by the U.S. Atomic Energy Commission.

Paul M. Chung
Lawrence Talbot
Kenell J. Touryan

Nomenclature

Numbers denote chapters in which symbol is used.

$\hat{a}$ defined in Eq. (3.39), **3**
a experimentally determined constant
A orbit of particle collected by probe (Fig. 2-4), **2**
A_p probe area, **2, 3**
$A^{\nu\nu}$ defined in Eq. (A.15), **App.**
b impact parameter, **App.**
$\hat{b}$ $d\bar{c}_e/dY$ concentration gradient, **4**
B orbit of reflected particle (Fig. 2-4), **2**
c mass fraction, **4**
characteristic speed $(kT_e/m_i)^{1/2}$, **1, 2, 3, App.**
C mass fraction, **3**
mass fraction, fluctuating quantity, **4**
$\vec{C}$ peculiar molecular velocity, **App.**
D diffusion coefficient
orbit that intersects probe but cannot be reached from infinity, Fig. 3-3, **3**
$\mathscr{D}$, $\mathscr{D}_e$ Damkohler numbers [Eq. (3.8a)], **3**
e electronic charge,
[see also Eq. (2.18), **2**]
E total energy [Eq. (2.6)], **2**
f Blasius function, **3**
$f(\ )$ probability distribution function, **App.**
[see also Eq. (2.16), **2**]
$F(\ )$ [see Eq. (2.16), **2**]
$F_0^{e\nu}$ a constant in intermolecular force law, **App.**
g relative speed between colliding particles, **App.**
g_s defined in Eq. (3.61), **3**
$G(E)$ curve in $E - \Omega$ space ($= \Omega^2$, angular momentum), **2**
G_α, $H_{\alpha\beta}$ functions defined in Eq. (A.37), **App.**
H total stagnation enthalpy per unit mass of plasma, **3**
H_e $h_e + \dfrac{e\varphi}{m_e}$, **3, App.**
h_e total enthalpy of electrons
I actual current collected by probe

j normalized current density
j^* $j\sqrt{\epsilon}$, **2**
J current per unit area
k Boltzmann's constant
K wave number, **4**
mobility, **App.**
K_h thermal conductivity, **App.**
K_{he} electron thermal conductivity, **3**
Kn Knudsen number
$K_{r,f}$ specific reaction rate, **3**
l probe length, **1, 2**
$\rho\mu/(\rho\mu)_0$, **3**
turbulence scale, **4**
L characteristic length
m particle mass
M ion-number density, fluctuating portion, **4**
Mach number, **1, 3**
$\mathcal{M}$ $c_i/c_{i\delta}$, **3**
n number density, **App.**
N mass fraction, **App.**
number density, **2, 3**
electron-number density, fluctuating portion, **4**
P, p total and partial pressure
Pr Prandtl number
$\vec{q}$ energy flux, Eq. (A.46), (A.84), **App.**
Q collision cross section
r radial distance
r_L gyroradius, **4**
R probe radius
characteristic length of body containing probe
Re Reynolds number, **1, 3, 4**
Rm magnetic Reynolds number, **4**
R_p characteristic length of probe electrode
R_s defined in Eqs. (3.17) and (3.19), **3**
s transformed coordinate, Eq. (3.40), **3**
Sc Schmidt number, **1, 3, 4**
S_j degree of ionization, Eq. (3.26), **3**
t time
T temperature, **2, 3, 4, App.**
(also trapped orbit, Fig. 2.3)
$\vec{u}$ mass averaged velocity, **3, App.**
$U_i(\ )$ potential energy, Eq. (2.4), **2**
u, v x and y direction velocity component, **3, 4**

U, V	x and y direction velocity fluctuating component, **4**
$\vec{V}$	velocity vector (U-component, **2**)
v_r, v_θ	radial and azimuthal velocity
$\dot{w}$	source term, mass rate per unit volume, Eqs. (3.1) and (3.2), **3**
$\dot{w}_{he}$	electron-energy source term, energy rate per unit volume, Eq. (3.3), **3**
x,y	axial and normal coordinates
Y	y/l, **4**
Z	charge number, **2, 3, App.**
α	exponent defined in Eq. (2.18), **2**
β	defined in Eq. (2.18), **2**
	$(m_e/m_i)^{1/2}$, **4**
	ratio of diffusion coefficients, Eq. (3.8a), **3**
Γ	charged particle flux, **3**
$\Gamma(\ )$	gamma function, **App.**
δ	boundary layer thickness, **3**
δ, ϵ	expansion parameters defined in Eqs. (A.13) and (A.11), respectively, **App.**
ϵ	T_i/T_e, **2**
	eddy diffusivity, **4**
$\bar{\epsilon}$	impact angle in two-particle collision, **App.**
ζ	shear stress, Eq. (A.84), **App.**
η	similarity variable, Eq. (3.47), **3**
θ	temperature ratio, Eq. (3.61), **3**
θ^ν	generalized function, **App.**
κ	a parameter, Eq. (3.29), **3**
λ_D	Debye length
λ_0	defined in Eq. (3.21), **3**
$\lambda_{\alpha\beta}$	mean free path for collision between species **1** and **2**
$\lambda^{\nu\bar{\nu}}$	mean free path for collision between species ν and $\bar{\nu}$, **App.**
λ_s	sheath thickness, **3**
μ	viscosity of plasma, **3, 4**
μ_0	permeability of free space, **4**
ν	kinematic viscosity
$\nu_{\alpha\beta}$	collision frequency between species **1** and **2**
ξ	r/λ_D, **2**
	separation distance, **4**
$\vec{\xi}$	absolute particle velocity vector, **App.**
ξ_p	r/λ_D, Debye ratio, **2**
ρ	gas density of mixture
σ	permittivity of free space
σ_c	electrical conductivity, **4**

τ	dimensionless time
	parameter defined in Eq. (2.28), **2**
	temperature ratio, Eq. (3.8a), **3, 4**
τ_l	parameter defined in Eq. (2.26), **2**
$\tau_{e,i}$	mean collision time, **4**
φ	electric potential
$\tilde{\varphi}, \tilde{\psi}$	perturbation functions, Eq. (A.16), **App.**
χ, χ^*	normalized potential, Eq. (2.3), **2, 3, 4**
χ_p	normalized probe potential, [Eq. (2.8a)] **2, 3, 4,**
ψ	nondimensional potential, Eq. (3.8a), **3**
$\omega_{e,i}$	electron and ion cyclotron frequency
ω_{pi}	ion plasma frequency, **2**
Ω	angular momentum, **2**

Subscripts

$a, i, e(\nu)$	neutral, ion, and electron, **3, 4, App.**
A	ambipolar, **3**
e, i	electron or ion
en, in	electron-neutral, ion-neutral
$+, -, e$	positive ion, negative ion, and electron, **4**
f	floating potential
n	neutral, **3**
o	conditions at reference point, or zeroth approximation, **3, 4**
p	probe, **2, 3**
r, R	radial and probe radius, **2**
s	sheath-edge conditions, **3**
sat	initial saturation point *a* of Fig. 1-3
t	turbulence
w	wall, **3**
α, β, γ	Cartesian tensor indices, **App.**
δ	boundary-layer edge conditions, **3**
μw	microwave, **4**
$\perp, \parallel$	perpendicular and parallel to magnetic field, respectively, **4**
∞	free-stream conditions

Superscripts

$a, i, e(\nu)$	atom, ion, and electron, **App.**
n, p, q	power indices, **3**
ϵ	exponent denoting two-dimensional (0) or axisymmetric (1) geometry, **3**
$\rightarrow$	vector notation
$\sim$	dimensionless quantity

( )′	value following collision, **App.**
	differentiation with respect to η, **3**
–	mean quantity, **4**
Δ( )	fluctuating quantity
*	reference conditions

Contents

Electric Probes in Stationary and Flowing Plasmas:
Theory and Application

CHAPTER I

Fundamental Considerations

The electric (or electrostatic) probe has long been used as a fundamental diagnostic tool for measuring the local properties of a plasma. In 1924 Langmuir pioneered the use of electric probes, which are often called Langmuir probes. Electric probes are relatively simple devices, but the theory underlying the probe response is, unfortunately, complicated. Basically, the electric probe consists of one or more small metallic electrodes inserted into a plasma. Two probe configurations are commonly employed. In the "single-probe" configuration, a single electrode is inserted into the plasma and attached to a power supply that can be biased at various potentials positive or negative in relation to the plasma; the current collected by the probe as a function of its potential is measured. The return electrode, which completes the circuit, is a "ground" for the plasma, and is typically a conducting portion of the wall of the vessel confining the plasma. In this case the probe potential is measured from the ground potential. The typical "double-probe" configuration consists of two electrodes, usually of equal area, that come in contact with the plasma, and the current passing through the plasma between the two electrodes is measured as a function of the voltage applied between them. Unlike the single probe, in the case of the double probe there is no net charge drain from the plasma, because the two electrodes and power supply of the double-probe system form an isolated closed circuit. Also, unlike the single probe, both electrodes of a double-probe system are always (in the case of an ion-electron plasma) negative with respect to the plasma. There are certain advantages associated with the use of each of the probe systems, as will be discussed subsequently.

Electric-probe theory is complicated because probes are boundaries to plasmas, and near boundaries the equations that govern the plasma behavior change. Charge neutrality does not hold near boundaries; a thin layer exists where electron- and ion-number densities differ, and the layer, often called a Debye sheath, can sustain large electric fields. In the absence of magnetic fields, the response of a probe in a flowing plasma depends on a number of parameters that arise naturally from the governing equations. These parameters may be combined into groups as indicated in Table 1.

The three parameters in Group I determine the various domains at which

Table 1. Grouping of Parameters

Group	Governing Parameters
I	λ_D/L; λ/L; λ/λ_D
II	χ_p; T_i/T_e; $\mathscr{D}$
III	Continuum probes: Sc_i; Re; β; M; T/T_w; γ
	Collisionless-transition probes: l/R; U/c

an electric probe can operate. As will become evident in our theoretical treatment in the Appendix, the various ranges of λ_D/R, λ/R,[1] and λ/λ_D determine the well-known domains of probe operation. These are collisionless/collisional limits; thin/thick sheath case; orbital limit, etc. These limits are illustrated in a three-dimensional R-λ_D-λ diagram shown in Fig. 1-1, using R as the characteristic probe dimension.

According to Fig. 1-1, for a given set of Group II and III parameters, one can identify six different probe regimes that could be divided into two domains depending on the magnitude of the Knudsen number, Kn ($Kn = \lambda/R$):

A. $Kn \gg 1$: Classical Langmuir Probe

1. $\lambda \gg R \gg \lambda_D$: conventional thin sheath

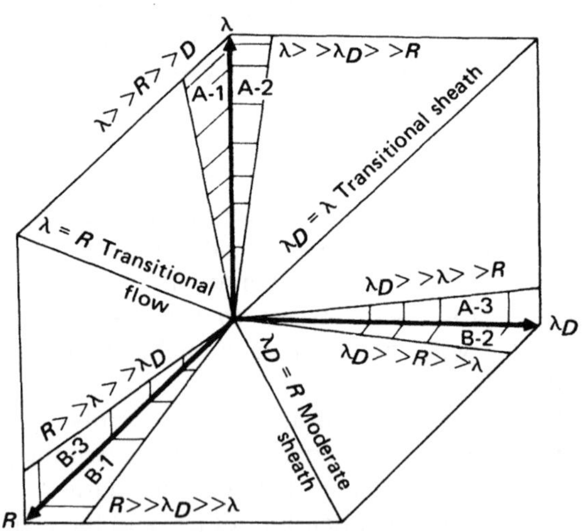

Fig. 1-1. Three-dimensional diagram illustrating various probe-operation regimes. R, probe radius; λ_D, sheath; λ, mean free path.

[1] For the moment, we use λ to denote a representative mean free path, without specifically defining the collision parameters. Also, for purposes of illustrating the nature of various domains of operation, we shall choose the characteristic length L to be the probe radius R.

2. $\lambda \gg \lambda_D \gg R$: orbital limit–thick sheath
3. $\lambda_D \gg \lambda \gg R$: collisional thick sheath (hybrid case)

B. $Kn \ll 1$: Continuum Electrostatic Probe

1. $R \gg \lambda_D \gg \lambda$: collisional thin sheath
2. $\lambda_D \gg R \gg \lambda$: collisional thick sheath
3. $R \gg \lambda \gg \lambda_D$: collisionless thin sheath (dense case)

Most probe theories are asymptotic in one of the three parameters, λ, λ_D, and R, and the domains of applicability reside in one of the shaded areas of Fig. 1-1. For example, regime A.1 lies in the $\lambda_D \simeq 0$ plane with λ as the largest parameter. Similarly, regime B.3 lies in the same plane but with R as the largest parameter.

Condition $Kn \gg 1$ does not necessarily represent the collisionless situation. The sheath may be collisionless (A-1, A-2) or collisional (A-3) depending on the relative magnitudes of λ and λ_D. The $Kn \ll 1$ represents the case of probes that are *always* collisional relative to the gas (continuum flow) but for which the sheath may or may not be collisional, depending on the relative magnitudes of λ and λ_D, viz., B-1 and B-2 as compared with B-3.

In addition one has to deal with several *transitional* domains, represented by the diagonals in Fig. 1-1. Between regions A-1 and B-3 the diagonal represents the $Kn \simeq 1$ transitional region, and between regions A-2 and A-3 there is the transitional sheath with $\lambda_D \simeq \lambda$. The transition from thin to thick sheath occurs both in the continuum (between B-1 and B-2) and in the collisionless (between A-1 and A-2) domains. Finally $R \simeq \lambda$ represents the double-transitional case.

Another way of representing the above domains is to exhibit them on a two-dimensional plot as in Fig. 1-2.

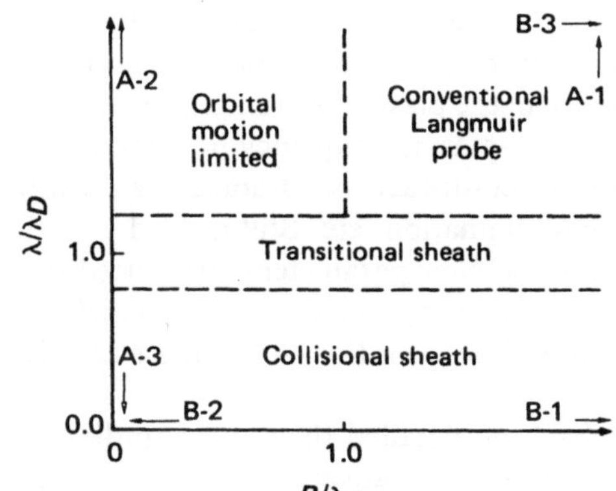

Fig. 1-2. Two-dimensional representation of various probe-operation regimes.

The most important parameter in Group II is the probe potential χ_p, where $\chi_p = e\varphi_p/kT_e$. It determines the current collection characteristics of the probe, and it may range from a large negative value to moderate positive values. The ratio of ion to electron temperature is given by T_i/T_e and for most plasmas it ranges from near zero to unity. It represents the level of thermal nonequilibrium in the plasma between electrons and heavy particles. $T_i/T_e = 1$ corresponds to the thermal equilibrium plasma and $T_i/T_e \to 0$ yields the cold ion with the frozen flow situation. The Damkohler number $\mathscr{D}$ is the ratio of the ionization–recombination rate to the ion-diffusion rate, and it represents the degree of chemical nonequilibrium, e.g., for $\mathscr{D} \to 0$, the flow is chemically frozen, and for $\mathscr{D} \to \infty$, the flow is in chemical equilibrium.

Group III contains parameters that depend on the plasma properties and particular probe geometry under consideration. In general, continuum probes commonly used consist of flush-mounted probes, blunt stagnation probes (e.g., hemisphere cylinders), or sharp cone probes. The collisionless-transition probes are usually cylinders or spheres and sometimes circular disks. For continuum probes the parameters Sc_i (ion Schmidt number), Re (Reynolds number), β (ion/electron-diffusion coefficient ratio), M (Mach number), T/T_w (plasma/probe-temperature ratio), and the specific heat ratio γ are usually the most important ones, but do not necessarily comprise a complete set. Their relative importance especially for thick, two-dimensional continuum sheaths is still under investigation.

The two additional parameters l/R and U/c listed under Group III appear specifically in the analysis of cylindrical probes, with axes aligned with the flow. Here, l/R is the aspect ratio of cylinder length to cylinder radius and U/c is the ratio of plasma-flow velocity to the characteristic speed $c = \sqrt{\dfrac{kT_e}{m_i}}$. They appear as probe end-effect parameters to correct for convective effects present in the flow.

Finally, for special situations, one should include several more groups of parameters in addition to the three mentioned above. One of these groups deals with turbulence in the plasma or transient phenomena, which introduces parameters such as ion-plasma frequency, ion-acoustic waves, etc.; another group deals with probe-surface conditions, e.g., emitting or nonemitting surfaces, surface contamination, etc. Still a third situation arises in the presence of magnetic fields. New parameters introduced in the latter include the electron and ion gyroradii (or gyrofrequencies), which combine with the other parameters to form additional ratios. All of these are discussed in some detail in Chapter IV.

Before we embark on detailed theoretical studies of probe operation, it would be well to review, qualitatively, the general behavior of a transitional or continuum probe immersed in an infinite plasma.

Figure 1-3 depicts the general shape of a current-voltage (CV) character-

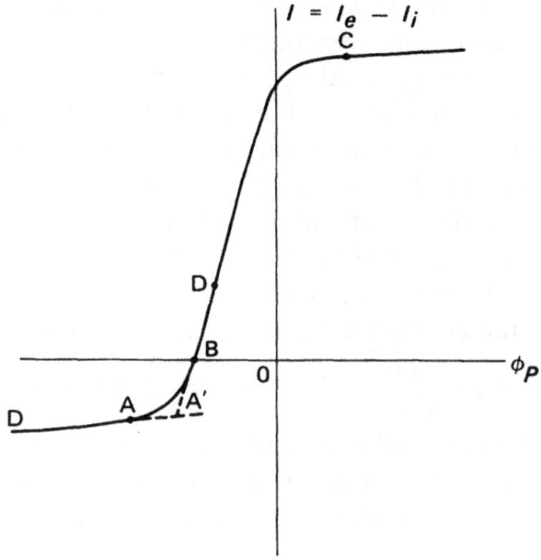

Fig. 1-3. Typical current-voltage (*CV*) characteristic curve for single-probe operation.

istic curve of a single probe in a typical plasma environment. This plot can be obtained by varying the voltage on the probe in a continuous fashion in a steady-state plasma or even in a transient plasma by the use of a fast-sweeping voltage source.

The qualitative behavior of the *CV* characteristic depicted in Fig. 1-3 can be explained as follows. For large negative values of the probe potential $\varphi_p < \varphi_{p,a}$, essentially all electrons in the vicinity of the probe are repelled from it. The electron current to the probe is negligible and the electric current to the probe consists of the ion current I_i, which is of the order of the natural ion-diffusion current. Although I_i will, in general, continue to increase for $\varphi_p \leq \varphi_{p,a}$, this branch of the characteristic is called the "ion-saturation" current, I_{is}. The nature of this "saturation" depends on the governing parameters discussed above, in particular, on whether the plasma is collisional or collisionless relative to the probe. In many cases, I_{is} can be conveniently related to the electron-number density N_e provided (T_e/T_i) is known.

When φ_p is made less negative relative to the plasma than its value at point *A*, the most energetic electrons in the plasma are able to overcome the retarding electric field of the probe and hence reach the probe, giving rise to an electron-current contribution to the measured probe current that decreases the net current observed. As the probe potential is made still less negative, a condition is reached at point B where the electron current collected exactly balances the ion current, and the net current to the probe is zero. This point is called the floating potential. At the floating potential, the probe is still negative with respect to the plasma, essentially collecting the saturation-ion current as well as a cancelling electron current.

In the region between $\varphi_{p,b}$ and $\varphi_{p,c}$, the probe potential is still negative, relative to the plasma potential $\varphi_{p,c}$, but the probe attracts increasingly more electrons so that the net current to the probe is an electron current. If the electron distribution were Maxwellian (as it often is), the shape of the curve here, after the ion contribution is subtracted, would be exponential and from the shape of this curve T_e could be determined.

Near $\varphi_p = \varphi_{p,c}$, the plasma potential, the electric field of the probe approaches zero and the electron current increases toward its natural diffusive value, which is of order $\sqrt{\frac{m_i T_e}{m_e T_i}}$ times the ion-saturation current. For $\varphi_p > \varphi_{p,c}$, the probe is at a potential that is positive relative to the plasma potential $\varphi_{p,c}$, and I increases slowly as ions are repelled and electrons are accelerated to the probe, giving rise to the "electron-saturation" phenomena, with a "saturation" current I_{es}.

The determination of the location of $\varphi_{p,c}$ on a CV characteristic where the probe is at the same potential as the plasma presents one of the more difficult problems of probe diagnostics. Very often there is no sharp change in slope between the electron-saturation region I_{es} and the region between $\varphi_{p,b}$ and $\varphi_{p,c}$ (often called electron-retarding region). When plotted on a semilogarithmic scale $\varphi_{p,c}$ can be located approximately by extrapolating the linear part of the characteristic to meet the best "straight" line obtained from the region $\varphi_p > \varphi_{p,c}$. [For further discussion on this point, see Swift and Schwar (1971).] A further difficulty in the determination of the plasma potential arises from the fact that the grounding of the plasma frequently is not ideal, in the sense that the plasma potential does not exactly coincide with ground potential, and a small positive probe potential relative to ground is required to obtain the natural electron-diffusion current. These preliminary remarks are given in a cautionary vein to indicate that probe measurements that rely crucially on an exact determination of the plasma potential are likely to be less reliable than those for which a modest uncertainty in its determination can be tolerated.

When there is thermal nonequilibrium in the plasma with $T_e > T_i$, the prediction of electron saturation current is much more difficult than the ion-saturation current. This is because T_e increases with positive φ_p as the field does work on the electrons. For this reason the portion of the CV characteristics where $\varphi_p < \varphi_{p,c}$ is generally the most useful for diagnostics in continuum plasmas.

The behavior of a double probe can be determined from a knowledge of the single-probe behavior of its components, and hence need not be treated separately. For example, suppose Fig. 1-3 represents the response of one of the two equal-area electrodes of a double probe when it is used as a single probe. Then in double-probe operation if one electrode were operating at point D on the characteristic, the other electrode would have to operate at point D′, such that $I_d = -I_{d'}$. In what follows, we will be concerned with single-probe response, unless otherwise noted.

The literature on probes is vast. Chen (1965) first summarized the available theoretical results in a systematic manner, supplementing them with practical information on experimental techniques. Since that time there has been an accelerated activity in the study of electric probes for arc-jet environments, on reentry vehicles and on satellites for measuring charged particle densities and electron temperatures.

The purpose of this volume is to make a systematic study of available probe theories and the use of electric probes in various plasma environments, with special emphasis on developments since 1965. Its ultimate aim is to arrive at useful expressions for interpreting the probe response to various plasma-operating regimes, and, wherever possible, to support the results with available experimental data.

Chapter II of this book is devoted to collisionless and transitional electric-probe behavior. Chapter III covers the continuum regime for probe operation. Engineering approximations are provided wherever possible and comparisons are made with available experimental data. Chapter IV deals briefly with special areas of probe operation where both analysis and experimental data are still scant. These include probe-surface phenomena; presence of negative ions; strongly ionized plasmas; plasmas in magnetic fields; and the response of collisionless and continuum probes to turbulence. Following Chapter IV is an Appendix in which the theoretical basis for probe analysis is developed starting from the kinetic theory of ionized gases. Various methods of solution are outlined briefly and domains of probe operation are more precisely delineated.

References

Chen, F. F. (1965), "Electric Probes," in *Plasma Diagnostic Techniques* (R. H. Huddlestone and S. L. Leonard, eds.), Academic Press, New York.

Swift, J. D. and Schwar, M. J. R. (1971), *Electric Probes for Plasma Diagnostics*, Iliffe Books, London.

CHAPTER II

Collisionless and Transitional Electric Probes

2.0 Introduction

In this chapter we will consider the response of electric probes operating in the collisionless and transitional regimes in both quiescent and flowing plasmas. In terms of the domains discussed in Chapter I we are concerned with roughly the region included and above the transitional sheath area demarcated in Fig. 1-2 of Chapter I. Our chief attention will be directed toward simple geometries, namely spherical and cylindrical probes, although mention of other geometries will be included.

Our first consideration will be that of probe response in quiescent (non-flowing) plasmas, in the collisionless limit where all relevant mean free paths are much larger than the characteristic probe dimension, which in the case of cylindrical and spherical probes is the probe radius R. For this collisionless limit, we examine how the probe response is influenced by the character of the surrounding sheath, as determined by the probe radius–Debye length ratio $\xi_p = R/\lambda_D$, which, for brevity, we shall call the "Debye ratio." Our main concern will be the relationship between the current-voltage (CV) characteristic of the probe and the plasma properties, principally the charged-particle density and the electron temperature; and we shall not concern ourselves with the details of sheath structure except insofar as it is necessary for interpreting the CV characteristic.

Following our examination of probe behavior in the collisionless limit we discuss the effect on probe response of collisions between charged particles and neutral particles or other charged particles, still for the case of a quiescent plasma. Here the relevant parameters, in addition to the Debye ratio, will be the Knudsen numbers $\lambda_{\alpha\beta}/R$. In considering the effect of collisions, we will be concerned mainly with the transition regime, as discussed in Chapter I, and not with the continuum limit, which is dealt with separately in the following chapter.

Finally, we discuss, to the extent that it is known, how the results obtained for probes in quiescent plasmas are modified when the plasma possesses a directed flow velocity. Attention will be focused mainly on cylindrical probes, and on the influence of an additional parameter, the aspect ratio l/R, where

l is the probe length.[2] Wherever possible, comparisons are made between theoretical and experimental results.

2.1 Early Theories

It will be useful at the outset to recall some of the early theories of probe response, particularly the pioneer work of Langmuir and his collaborators. Excellent and quite complete accounts of this material have been given by Chen (1965a) and by Swift and Schwar (1971), so we shall not attempt to be exhaustive here, but rather shall confine ourselves to the mention of those results that are most important for our subsequent discussions of the more recent theories. Although the early theories have for the most part been superseded by the more recent ones, in some instances they represent valid limiting cases of the more general results, and are therefore still of interest.

For the sake of definiteness, we shall generally assume that the ions are singly charged, and are the species attracted to the probe, although the results we present are equally applicable to electron-attracting probes, and can be scaled so as to apply to multiply charged ions. We define the normalized current densities j_i and j_e collected by the probe according to the relationship

$$I_{i,e} = A_p N_\infty Z_{i,e} e \sqrt{\frac{kT_{i,e}}{2\pi m_{i,e}}}\, j_{i,e} \tag{2.1}$$

Usually, $Z_i = 1$, although cases can arise involving multiply-charged ions or negative ions, which take the place of electrons in maintaining charge neutrality. We note that the factors $J_{R(i,e)} = N_\infty Z_{i,e} e(kT_{i,e}/2\pi m_{i,e})^{1/2}$ are the random thermal ion and electron currents collected by a probe at the plasma potential in a Maxwellian plasma so that the normalization consists of referring the probe current densities $I_{i,e}/A_p$ to these random currents is

$$j_{i,e} = (I_{i,e}/A_p)/J_{R(i,e)} \tag{2.2}$$

However, it is often more convenient to normalize the probe ion current with respect to the random ion current evaluated at the electron temperature, in which case we denote the normalized ion current j_i^*, and observe that $j_i^* = j_i\sqrt{\epsilon}$, where $\epsilon = T_i/T_e$. The reason this normalization is particularly useful in that it relates back to the so-called "Bohm condition" (Bohm *et al.* 1949),

[2] Actually, the aspect ratio l/R is a relevant parameter for cylindrical probes in quiescent plasmas as well, because even the "longest" cylindrical probes with $l/R \gg 1$ are three-dimensional objects and experience some current contributions from the ends. However, if $l/R \gg 1$, the current contribution from the ends of the probe will generally be negligible, unless the probe potential is so large that the sheath thickness becomes comparable to the probe length. Some discussion of this point is given by Laframboise and Parker (1973). In what follows regarding cylindrical probes in quiescent plasmas, it is assumed that l/R is sufficiently large that end effects are negligible.

which establishes the result that for an ion-attracting probe, electrical fields in the quasi-neutral region exterior to the sheath accelerate ions such that they enter the sheath with approximately the velocity $\left(\frac{kT_e}{m_i}\right)^{1/2}$. Therefore the random ion current evaluated at the electron temperature can be expected to provide a better reference for the actual ion current than a random current evaluated at the ion temperature.

We will use two normalized potentials for ion collection, generally given by

$$\chi = Ze\varphi/kT_i \tag{2.3a}$$

$$\chi^* = Ze\varphi/kT_e \tag{2.3b}$$

where $Z = Z_i$ (we have suppressed the subscript i) and φ is the local potential measured relative to the plasma potential. When dealing with electron-current collection, we will also use the potential χ^*, with $Z = 1$. Although the inclusion of the factor Z makes the analysis valid for multiply charged ions, most practical cases involve singly charged positive ions, for which $Z = 1$. Therefore, to avoid unnecessary confusion, unless Z is specifically exhibited in an equation, it should be understood that $Z = 1$ and that the singly ionized particles are under consideration. If, as is rarely the case, the situation $Z \neq 1$ must be considered, the necessary modifications in the definitions of χ and χ^* can easily be made.

2.2 Orbital Motion Limit

One of the most important early results concerns current collection by spherical and cylindrical probes in the *orbital motion limit* (OML), a terminology introduced by Langmuir and Mott-Smith (1926), who gave the results for this limit for charged particles having both monoenergetic and Maxwellian velocity distributions. The OML current is that current collected by a probe when none of the undisturbed plasma particles (at infinity) capable of reaching the probe on the basis of energy considerations is excluded from doing so by intervening barriers of effective potential. For the attracted particles, the OML corresponds to the limit $R/\lambda_D \rightarrow 0$, that is, the infinitely thick sheath limit (see Appendix, Section A.2.1).

As is well known [cf. Goldstein (1950)], the two-body central-force problem can be reduced to an equivalent one-dimensional problem in which only the radial velocity v_r of the particle is considered (see Fig. 2-1) if the local potential $\varphi(r)$ is replaced by an effective potential, for the case of ions,

$$U_i(r,\Omega) = Ze\varphi(r) + \frac{\Omega^2}{2m_i r^2} \tag{2.4}$$

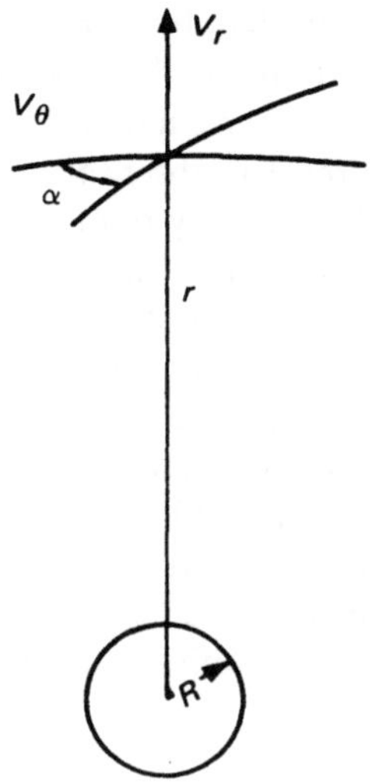

Fig. 2-1. Coordinate system for spherical probe.

where Ze is the charge on the ion, and Ω is the angular momentum of the ion given by

$$\Omega = m_i r v_\theta \tag{2.5}$$

and is an invariant of the motion. Because the total energy of a particle is

$$E = \tfrac{1}{2}m_i(v_r^2 + v_\theta^2) + Ze\varphi(r) = \tfrac{1}{2}m_i v_r^2 + U_i(r,\Omega) \tag{2.6}$$

and this total energy is an invariant of its motion, it follows that a particle starting from infinity with particular values of E and Ω will reach a particular radius r only if $E - U_i(r) \geq 0$; otherwise v_r would be imaginary. Now, the particular feature of the OML is that if at a particular radius r_1 the condition $E - U_i(r_1) \geq 0$ is satisfied, for a particular particle, then it is also satisfied for that particle for all $r > r_1$. In other words, there are no locations $r > r_1$ for which the effective potential $U_i(r)$ has a local maximum where $E - U_i(r) < 0$. This would act as a potential barrier to reflect particles back from their inward motion, even though the particles had sufficient energy to satisfy the condition $E - U_i(r_1) \geq 0$ at an inner radius, which indeed could be the radius of the probe. Of course it is not obvious under what conditions $U_i(r)$ will not have a local maximum, since it is a combination of a positive term $\Omega^2/2m_i r^2$ and a term $Ze\varphi(r)$, which, in the case of an attracting probe, will be negative, the magnitude of which must be determined by a self-consistent solution of Poisson's equation. This, in fact, is the crux of the difficulty involved in the exact solution of the general collisionless probe problem.

However, if it is assumed a priori that potential barriers do not exist, the calculation of the current collected by a probe involves only energy and angular momenta considerations, and does not require a simultaneous solution of Poisson's equation. It can, in fact, be shown for ion-current collection that the condition required for absence of a local maximum in $E - U_i$ is that $|\varphi(r)|$ must decrease less rapidly than r^{-2} with increasing r.

The detailed calculations for orbital-motion-limited current collection by spherical and cylindrical probes was first carried out by Langmuir and Mott-Smith (1926). These calculations were performed for a sheath of finite dimension, and the limit for an infinite sheath was taken as a special case. However, subsequent analyses have established that, at least for a sphere, only in the infinite sheath limit ($R/\lambda_D \to 0$) are the conditions necessary for the OML theory obtained, so we shall give only the infinite-sheath results.

Considering ions to be the attracted species ($\chi_p < 0$), we have the following results for the orbital-motion-limited current, for a Maxwellian distribution at infinity.

For the spherical probe

$$j_i = 1 - \chi_p \tag{2.7a}$$

$$j_e = e^{\chi_p^*} \tag{2.7b}$$

and for the cylindrical probe

$$j_i = \frac{2}{\sqrt{\pi}}\left\{\sqrt{-\chi_p} + \frac{\sqrt{\pi}}{2}e^{-\chi_p}[1 - \operatorname{erf}(\sqrt{-\chi_p})]\right\} \tag{2.8a}$$

$$j_e = e^{\chi_p^*} \tag{2.8b}$$

All these expressions apply as well for the case in which electrons are the attracted species, if interchange is made between subscripts i and e and the factors $-\chi_p$ and χ_p^*.

The analysis of probe behavior is simplified if it is assumed that the attracted species have a monoenergetic velocity distribution, rather than the more realistic Maxwellian one. If the monoenergetic velocity distribution is chosen such that the random thermal currents for the Maxwellian and monoenergetic cases are the same at a given temperature, then it is found that for the sphere the monoenergetic ion energy E_M should be taken to be $E_{M_{\text{sph}}} = (4/\pi)kT_i$, and for a cylinder $E_{M_{\text{cyl}}} = (\pi/4)kT_i$. For orbital-motion-limited ion-current collection in the monoenergetic ion approximation, the result for the sphere is

$$j_i = 1 - \frac{\pi}{4}\chi_p \qquad \left(\chi_p \le \frac{4}{\pi}\right) \tag{2.9a}$$

$$j_e = e^{\chi_p^*} \tag{2.9b}$$

and for the cylinder it is

$$j_i = \sqrt{1 - \frac{4}{\pi}\chi_p} \quad \left(\chi_p \leq \frac{\pi}{4}\right) \tag{2.10a}$$

$$j_e = e^{\chi_p^*} \tag{2.10b}$$

These monoenergetic results were also given by Langmuir and Mott-Smith, although they did not specify the relationship between E_M and T_i.

It was mentioned earlier that the orbital-motion-limited current is the current collected by a probe in the limit $R/\lambda_D \to 0$. Detailed calculations to be described later show that this is a more precise statement for the case of a spherical probe than for a cylindrical probe. For the latter there appears to be a domain between $R/\lambda_D = 0$ and a small but finite value of order unity over which the orbital-motion-limited current is obtained. For the sphere, the OML current is reached only in the limit $R/\lambda_D \to 0$.

Laframboise and Parker (1973) have reexamined the orbital-motion-limited regime, and rederived the results given above on the basis of energy considerations alone. In so doing they are able to establish that the results apply not only to circular cylinders and spheres, but to any convex cylindrical shape and to certain classes of sufficiently convex three-dimensional shapes. In this connection it is of interest to note, as can be ascertained from Eqs. (2.7a) and (2.8a), or from Eqs. (2.9a) and (2.10a), that in the OML the slope of the current-voltage characteristics of a spherical probe is proportional to $N_\infty/\sqrt{T}$, where T is the temperature of the attracted species, whereas for a cylindrical probe at large probe potentials the slope is proportional to N_∞. Therefore if both a spherical and a cylindrical probe are operated simultaneously it should, in principle, be possible to determine N_∞, T_i, and T_e for the plasma, T_e being found either in the classical fashion from the slope of the retarding field portion of the characteristic or by choosing electrons as the attracted species. As mentioned previously the difficulty in achieving this in practice arises because the OML for a spherical probe is reached asymptotically only as $\xi_p \to 0$, which implies the requirement of a very tenuous plasma ($N_\infty < 10^6$ cm^{-3}) for a probe of practical size. Laframboise and Parker give a very useful discussion of this problem, and suggest some strategies for circumventing it through the use of probe geometries that give spherical-probe behavior but are capable of greater miniaturization than is the sphere itself.

However, even if the conditions in a plasma do not correspond to the OML regime for a given probe, the OML results are still of importance in that they provide an upper bound for the current collected by a probe under collisionless conditions. This is because potential barriers, which occur at finite values of R/λ_D, can only reduce the number of charged particles which are able to reach the probe.

2.3 The Cold-Ion Approximation

Because in many plasmas $T_i/T_e \ll 1$, it is natural to examine the nature of probe response in the limit $T_i/T_e = 0$. This was done by Allen *et al.* (1957) for the case of a spherical probe in a collisionless plasma. They started with the equations of Bohm *et al.* (1949), which are the same as the equations later used by Bernstein and Rabinowitz (1959), and which apply for monoenergetic ions. In the case of an ion-attracting spherical probe Allen *et al.* show that it is correct to assume that in the limit $T_i/T_e = 0$ the ions move in a radially inward direction, and that their velocity arises solely from the energy they acquire in the potential field of the probe. Thus their velocity is given by

$$v_r = \left(-\frac{2e\varphi}{m_i}\right)^{1/2} = \left(\frac{2kT_e}{m_i}\right)^{1/2}(-\chi^*)^{1/2} = v_s(-\chi^*)^{1/2} \qquad (2.11)$$

and with the current to the probe being $I_i = 4\pi er^2 N_i v_r$, the ion density by continuity is

$$N_i = I_i/[4\pi er^2 v_s(-\chi^*)^{1/2}] \qquad (2.12)$$

If the electrons are in a Boltzmann distribution, such that $N_e = N_\infty e^{\chi^*}$, then Poisson's equation can be written

$$\frac{1}{r^2}\frac{d}{dr}\left(r^2\frac{d\varphi}{dr}\right) = -\frac{e}{\sigma}\left(\frac{I_k}{4\pi r^2 ev_s(-\chi^*)^{1/2}} - N_\infty e^{\chi^*}\right) \qquad (2.13)$$

or with $\xi = r/\lambda_D$, $\xi_p = R/\lambda_D$, and $I_i = 4\pi R^2 eN_\infty(kT_e/2\pi m_i)^{1/2} j_i^*$

$$\left(\frac{d^2\chi^*}{d\xi^2} + \frac{2}{\xi}\frac{d\chi^*}{d\xi} - e^{\chi^*}\right)(-\chi^*)^{1/2}\xi^2 = I_i - \frac{j_i^*\xi_p^2}{2\sqrt{\pi}} \qquad (2.14)$$

This is the equation solved by Allen *et al.* Some results are given in their paper, and more complete results have been computed by Chen (1965b). The most extensive investigations of Eq. (2.14) are those carried out by Laframboise (1966) and are plotted in Fig. 2-2(a).

The importance of the ABR calculation resides not only in the utility of the results for practical applications, but also because the ABR equation represents the correct limit for $T_i/T_e \to 0$ of the exact equations for a spherical probe, both for the monoenergetic ion approximation treated by Bernstein and Rabinowitz (1959) and the Maxwellian ion distribution considered by Laframboise (1966). The ABR analysis is particularly important for the Laframboise calculations, because the nature of Laframboise's numerical scheme prevented him from carrying out calculations in the limit $T_i/T_e = 0$; thus the ABR values provided the necessary end points for the effect of varying T_i/T_e on the ion-current collection by a spherical probe.

An interesting point arises with regard to the cold-ion ($T_i/T_e \to 0$)

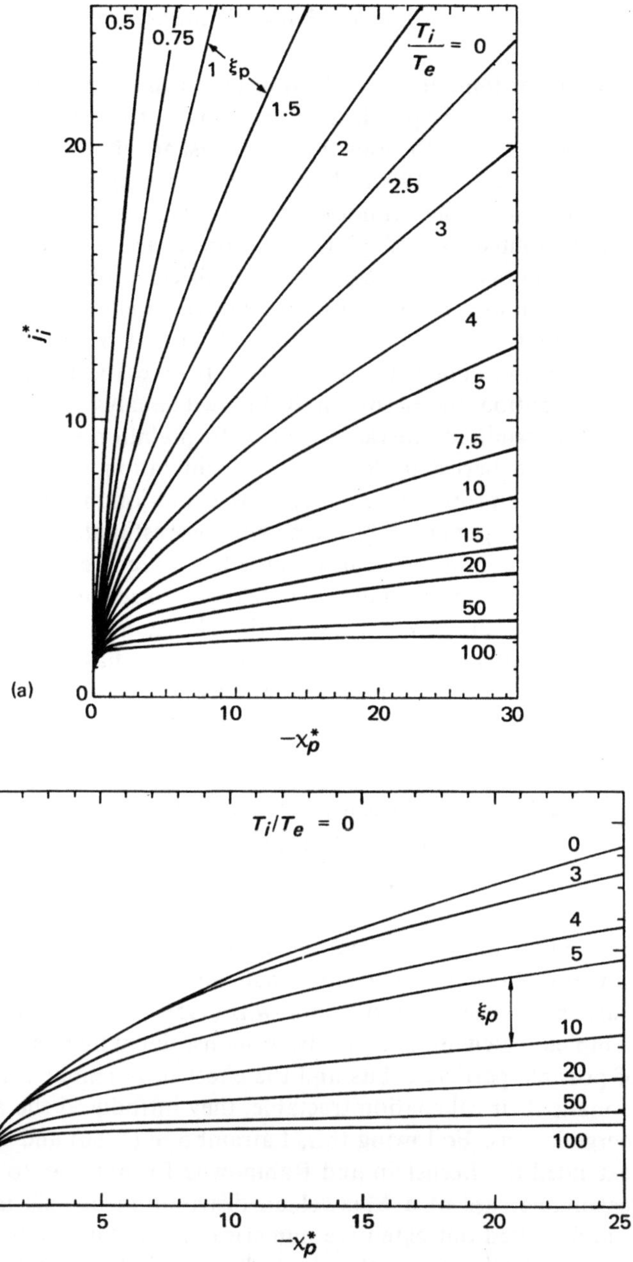

Fig. 2-2. (a) Ion current j_i^*-versus-probe potential for various ratios of probe radius to electron Debye length; ion-attracting spherical probe; $\epsilon = 0$; obtained from numerical solution of the Allen–Boyd–Reynolds (ABR) equation. (From Laframboise, 1966.) (b) Ion current j_i^*-versus-probe potential for various ratios of probe radius to electron Debye length; ion-attracting cylindrical probe; $\epsilon = 0$. (From Laframboise, 1966.)

approximation for ion-current collection by a cylindrical probe. Allen *et al.* showed that, unlike the spherical probe, it is not correct in the limit $\epsilon = T_i/T_e \to 0$ to assume that the ions start from rest at infinity and move radially inward. However, Chen (1965b) later used the radial motion assumption to derive an equation for the cylindrical probe analogous to the ABR equation, and from it computed a set of cylindrical probe characteristics. The reason for this apparent anomaly with regard to cylindrical probe behavior has been given by Laframboise (1966) and Lam (1965a). Briefly, it has to do with the fact that the limit $E_i \to 0$ is a singular limit, in that the angular momentum Ω_i becomes indeterminate as $r \to \infty$. If $\Omega_i(r \to \infty)$ is allowed to have an isotropic distribution, the Bernstein–Rabinowitz analysis results. If, however, only $\Omega_i = 0$ is permitted, the radial motion formulation ensues. It is generally accepted that the Bernstein–Rabinowitz formulation, which allows for all values of Ω_i, is the proper one; therefore the cold-ion model, which does not take into account potential barriers, will overestimate the current under truly collisionless conditions. Laframboise has, in fact, calculated the current to a cylindrical probe in the limit $\epsilon = 0$, taking into account potential barriers, the results of which are given in Fig. 2-2(b). However, it also appears that the radial motion results of Chen for the cylinder may have application to ion-current collection under the special situation where ion–ion collisions become important. This point is discussed in Section 2.5.

2.4 Exact Theories for Current Collection by Spherical and Cylindrical Probes in the Collisionless Limit

There are two analyses for current collection by spherical and cylindrical probes under collisionless conditions that can be considered to be exact. One is the analysis of Bernstein and Rabinowitz (1959), who were the first to carry out a complete calculation of probe response including essentially the full range of possible particle orbits and the effect of potential barriers (see Fig. 2-3). To make their calculation tractable, they introduced the assumption of monoenergetic ions. Following this, Laframboise (1966) and Hall and Fries (1966) extended the Bernstein and Rabinowitz formulation to the physically more reasonable case of a Maxwellian distribution of ions in the ambient plasma, and carried out extensive numerical computations of ion- and electron-current collection for both spherical and cylindrical probes, over a wide range of Debye ratio R/λ_D, temperature ratio T_i/T_e, and probe potential χ_p. The calculations of Laframboise and of Hall and Fries are sufficient for establishing the current-voltage characteristics of spherical and cylindrical probes over essentially the entire range of practical conditions of operation, in the collisionless limit. It is not possible in the space allotted here to give an

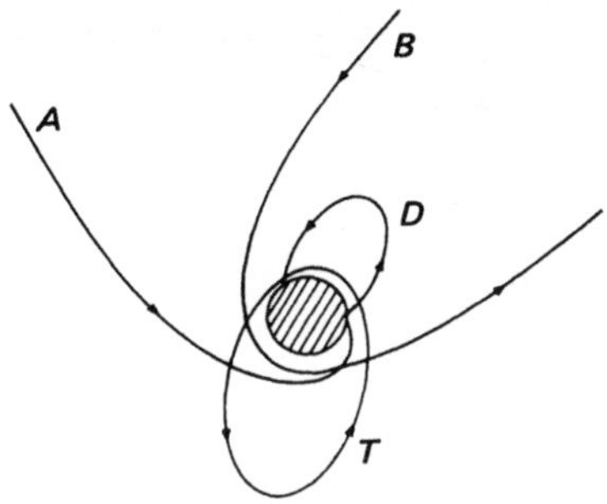

Fig. 2-3. Schematic of the various types of orbits. *A* represents an orbit of a particle collected by the probe. *B* is an orbit of a particle that encounters a potential barrier or turning point, and is reflected back into the plasma. *T* represents a trapped orbit (not necessarily closed) and *D* represents an orbit intersecting the probe, which is not attainable from infinity.

account of the analysis and computational methods employed, but we shall exhibit some of the results obtained by Laframboise.

Some typical results obtained by Laframboise for ion and electron collection by cylindrical probes at a probe potential $|\chi_p^*| = 25$, for several values of ϵ, are shown in Figs. 2-4(a)(b). Several things may be noted in these figures. For ion collection, Fig. 2-4(a), the break in the curves for the monoenergetic cases occurs at the onset of the OML, and for all ξ_p less than the value at the break point, the current is OML and constant at a value that depends only very slightly on ϵ, if $|\chi_p^*| > 1$. The current in the Maxwelliam case approaches the OML smoothly rather than discontinuously, but in the limit $\xi_p \to 0$ agrees quite well with the monoenergetic case. When the current is not OML, in the range $\xi_p \gtrsim 5$, the monoenergetic and Maxwellian results also agree quite closely, giving support to the usefulness of the Bernstein–Rabinowitz monoenergetic approximation. In fact it is only in the region of transition to the OML that the monoenergetic and Maxwellian results differ appreciably.

The results for electron-current collection, shown in Fig. 2-4(b), exhibit essentially the same features as those for ion collection. Again, the monoenergetic and Maxwellian cases are in good agreement for $\xi_p \gtrsim 5$. However, the discrepancy between the two is greater than for ion collection for $\epsilon < 1$ in the transition region to the OML, and the OML is approached more slowly than in the case of ion collection. This is associated with the fact that the collected species here is at a higher temperature than is the repelled species.

Laframboise's results for the spherical probe are for the most part qualitatively similar to those shown here for the cylindrical probe. One difference between the two cases is that the OML is attained at a finite value of ξ_p for the cylindrical probe, but it appears that for the spherical probe the OML is reached only in the limit $\xi_p = 0$ for Maxwellian particles. In addition,

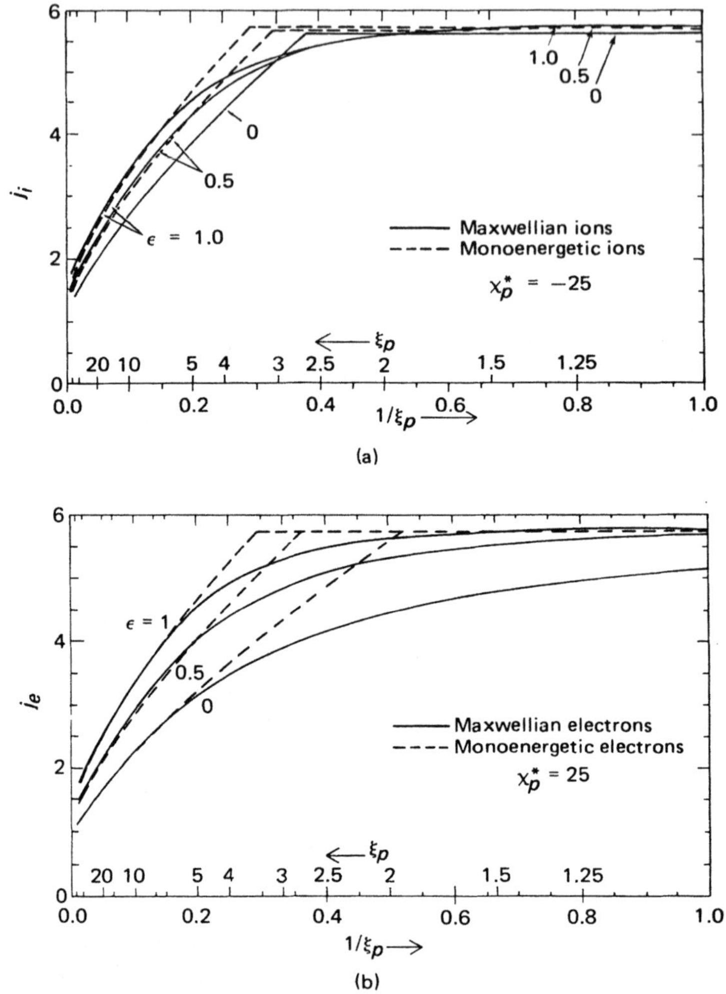

Fig. 2-4. (a) Ion current to cylindrical probe in a quiescent plasma as a function of ξ_p for $\chi_p^* = -25$, for various values of ϵ. (From Laframboise, 1966.) (b) Electron current to cylindrical probe in a quiescent plasma as a function of ξ_p for $\chi_p^* = 25$, for various values of ϵ. (From Laframboise, 1966.)

the monoenergetic and Maxwellian results are not in as good agreement in the limit $\xi_p \to 0$ as they are for the cylindrical case.

Several approximate fits to Laframboise's results have been made. One by Kiel (1968) takes the following form for ion collection by a cylindrical probe, with $Z_i = 1$,

$$j_i^* = F(\epsilon)\left[1 + \frac{f(\epsilon)}{\xi_p^{3/4}}(-\chi_p^*)^{1/2}\right] \tag{2.15}$$

where

$$F(\epsilon) = \epsilon^{1/2}(e^{\chi_s} \operatorname{erfc} \sqrt{\chi_s} + 2\sqrt{\chi_s/\pi})$$
$$\chi_s = 0.693/\epsilon \quad \text{for} \quad \epsilon \leq 1 \tag{2.16}$$
$$f(\epsilon) = 2.18(1 - 0.2\epsilon^{0.35})(1 + \epsilon)^{-1/8}$$

A slightly different form of the equation can be found in the paper for the case of electron collection, with $\epsilon \geq 1$, and also modifications for $Z_i > 1$ are given. Kiel (1971) has also developed a fitting formula for Laframboise's results for the spherical probe.

Another fitting formula for the cylindrical probe has been developed by Peterson and Talbot (1970), which takes the form for both ion- and electron-current collection

$$j^*_{i,e} = (\beta + |\chi^*_p|)^\alpha \tag{2.17}$$

with

$$\alpha = a/(ln\, \xi_p + b) + c\epsilon^m + d$$
$$\beta = e + \epsilon\{f + g(ln\, \xi_p)^3 - l/\xi_p\} + ln\, \xi_p \tag{2.18}$$

and the constants a, b, c, d, e, f, g, l, and m have the values (assuming that $\epsilon \leq 1$) given in the following table.

Table 2. Values of Constants in Eqs. (2.17)–(2.18)

	a	b	c	d	e	f	g	l	m
Ion collection	2.900	2.300	0.070	−0.340	1.500	0.850	0.135	0.000	0.750
Electron collection	2.900	2.300	0.110	−0.380	−2.800	5.100	0.135	2.800	0.650

The Kiel and Peterson–Talbot fitting formulas for the cylindrical probe are about of equal accuracy, agreeing with Laframboise's numerical results generally within less than 3 per cent error. Kiel's formula for the spherical probe is about of the same accuracy, although there is some difficulty in approximating the cold-ion limit $\epsilon \ll 1$, which is the ABR case. None of the fitting formulas is accurate in the OML, so they should be used only for $\xi_p \gtrsim 5$. The advantages of these fitting formulas, where they apply, are twofold. The obvious advantage is that they provide a means for interpolation and extrapolation of the numerical results given by Laframboise. The other is that they provide analytical representations of the ion and electron current, which can be used to express the characteristics of double-probe response for finite ξ_p, when the "saturation" currents are voltage-dependent, thus extending the work of Johnson and Malter (1950), who assumed that the saturation currents were constant, independent of probe potential. This extension for the cylindrical probe is discussed in detail by Peterson and Talbot (1970).

A general comment may be made regarding the use of Laframboise's results or the approximations thereto. It will be noted that in order to apply them one must know both $\epsilon = T_i/T_e$ and ξ_p. Generally, one can determine T_e from the electron-retarding region of the probe characteristic, if suitable care is taken [cf. Kirchhoff *et al.* (1971)], but in general T_i must be obtained or estimated by some independent means, such as a gas-dynamic measurement or calculation. (An exception to this, related to transient probe response, is discussed later.) However, in many practical circumstances, $\epsilon \ll 1$, and the results for the $\epsilon = 0$ limit can be used with some confidence. Because the Debye ratio ξ_p involves both T_e and the ambient charged particle density N_∞, which is usually the quantity to be determined, it too is initially unknown, and it would seem that some iteration would be required. However, Sonin (1966) has shown how iteration can be avoided if one wishes to determine N_∞ from, say, a single ion-current measurement, assuming that the value of ϵ has been established by some means. Sonin suggests that one choose some fixed potential for the probe, one that is sufficiently negative for electron current to be negligible, e.g., 10 dimensionless volts below the floating potential, so that $\chi_p^* = \chi_f^* - 10$. (χ_f^* itself is of the order of -5.)[3] He notes that the quantity

$$j_i^* \xi^2 = (R^2/\sigma)(2m_i/e)^{1/2}(e/kT_e)^{3/2}(I_i/A_p) \tag{2.19}$$

is a function only of T_e and the measured probe current density (I_i/A_p), even though j_i^* itself is a function of ξ_p, ϵ, and χ_p^*. Sonin therefore suggests that for a chosen fixed probe potential, say $\chi_p^* = \chi_f^* - 10$, one constructs curves of j_i^*-versus-$j_i^* \xi_p^2$ for various values of ϵ. Because $j_i^* \xi_p^2$ is known from the measurement, j_i^* can be read from such a plot and thus ξ_p^2 can be determined (see Fig. 2-6). Of course once ξ_p is known, so is N_∞. This procedure has been used with satisfaction by many workers. The procedure implies that T_e has been determined by some means, presumably from the slope of the retarding-field portion of the probe characteristic. Therefore, even if only one point on the ion-current curve is used to determine $j_i^* \xi_p^2$ the full characteristic usually will be required. Obtaining the full probe characteristic whenever possible is good practice in any case; otherwise possible anomalies in probe response may go undetected.

We have thus far made no mention of the several analytical studies that

[3] The floating potential χ_f^* can be determined by equating the electron current in the electron-retarding region, $j_e^* = \exp(\chi_p^*)$, to the product $\sqrt{\dfrac{m_e}{m_i}}\, j_i^*$ with j_i^* given numerically by Laframboise, or alternatively as expressed by one of the approximating formulas, Eq. (2.15) or Eq. (2.17).

have been carried out for cylindrical and spherical probes, in particular the asymptotic analysis of Lam (1965a,b). Lam's analysis, which was carried out before Laframboise's numerical results became available, is based on the Bernstein–Rabinowitz monoenergetic ion model, and involves the use of matched asymptotic expansions, in the limit $\xi_p \gg 1$, for the quasi-neutral, intermediate, and sheath regions surrounding spherical and cylindrical probes. This analysis is quite elegant and provides much insight regarding the detailed structure of the several regions mentioned, but for the practical purpose of constructing a current-voltage characteristic it is now better to use the Laframboise results (or the fitting formulas constructed for them) than to use the results of Lam's earlier asymptotic analysis. An exception to this might be found if it were desired to examine the detailed nature of the approach to the $\xi_p = \infty$ limit, because this is the region where Lam's analysis is most accurate, and also where (for $\xi_p < 100$) essentially no numerical results exist. Alternatively, one may wish to use the results of the Lam analysis for values of $|\chi_p^*|$ in excess of 25, the limit of the Laframboise calculations, in preference to the fitting formulas.

Before leaving the subject of probe response in the collisionless regime, it is appropriate to say a few words about experimental verification of the theory. There have been several attempts to verify the Laframboise results, in both quiescent and flowing plasmas. In the case of a quiescent plasma, one of the most satisfactory verifications of the theory is that which was carried out by Chen *et al.* (1968). Chen *et al.* tested both cylindrical and spherical probes in a potassium plasma and compared the inferred charged-particle concentrations with those determined simultaneously by microwave measurements. Because of the nature of the experiment, the cylindrical probes were operated mainly in the orbital-motion regime, but the spherical probes were operated in a regime that spanned a considerable range of ξ_p, from a value low enough for the current to approach the orbital limit at $\xi_p = 0$ (for Maxwellian ions) to a value of $\xi_p = 30$, where considerable departure from the OML current is found. For both types of probes good agreement with the Laframboise results were found over the entire range of operation. The results of Chen *et al.* (1968) for spherical probes are shown in Fig. 2-5. As can be seen from this figure, the Lam analysis was found also to give accurate predictions in the range of its validity, which is $\chi_p \xi_p^{-4/3} = O(1)$, with $\xi_p^2 \gg 1$.

Attempts have also been made to verify Laframboise's results for cylindrical probes in flowing plasmas, under the assumption that a cylindrical probe aligned with the flow direction will, if all relevant mean free paths are significantly greater than the probe radius, respond in the same fashion as a cylindrical probe in a stationary plasma. Under certain conditions this assumption has proved to be correct and results in agreement with the Laframboise calculations have been found, as for example by Sonin (1966), Dunn and Lordi (1970), and by Lederman *et al.* (1968). However, there are circumstances in

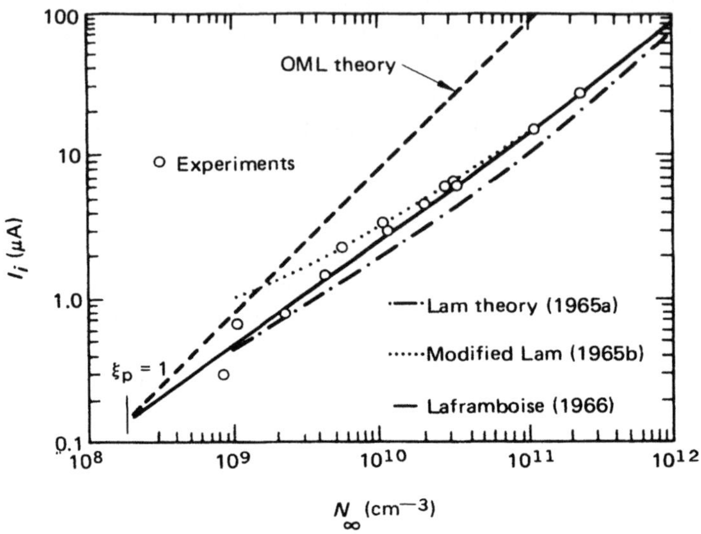

Fig. 2-5. Measured values of ion current for a spherical probe as a function of plasma density. Probe radius = 0.025 cm, $\chi_p^* = -20$. Plasma density determined by microwaves. (From Chen *et al.*, 1968.)

which this assumption does not hold. These are discussed in Section 2.6.

In the case of Sonin's measurements in a flowing plasma, as will be discussed in more detail later, apparent agreement with Laframboise's calculations was found except in the orbital-motion-regime, but it was subsequently determined that *ion–ion collisions* played an important role in the measurements. Dunn and Lordi found good agreement between their data and the Laframboise results in a small range of the orbital-motion-regime where $\xi_p \gtrsim 2$, but found that the measured currents exceeded the theoretical values for lower values of ξ_p. Somewhat similar results were obtained by Lederman *et al.* However, both the Dunn and Lordi and the Lederman *et al.* results were subsequently determined to contain an *end-effect* contribution, about which more will be said later. The resolution of these ion–ion and end-effect contributions has eventually led to substantiation of the use of the Laframboise results for aligned cylindrical probes in flowing collisionless plasmas, as well as in stationary plasmas. Further discussions of the end-effect will be given in Section 26.

The Laframboise results for electron current collection by cylinders also have been substantiated by experiment. Dunn (1972) found excellent agreement between number densities obtained by ion-current-collecting and electron-current-collecting cylindrical probes, for $\xi_p > 1$, in both shock tunnel, free-stream and flat-plate boundary-layer flows.

2.5 Collisional Effects on Probe Response

When the ion mean free paths for collisions with either neutral gas atoms or with other ions are not large compared to the probe radius, the collisionless theory discussed in the previous section no longer applies and the solution of the Boltzmann and Poisson equations must be obtained as discussed in the Appendix. Several attempts have been made along these lines to analyze in particular spherical probe response under the influence of ion-neutral collisions, in a stationary plasma. One of the earliest of these is the work of Wasserstrom *et al.* (1965), and this was later followed by the studies of Chou *et al.* (1966), Self and Shih (1968), and Bienkowski and Chang (1968). From an analytical point of view, the kinetic theory approach of Chou, Talbot, and Willis (CTW) is perhaps the most rigorous, although the numerical work necessary to generate a useful family of probe characteristics is prohibitive.

The CTW formulation is based on a moment method solution of the Boltzmann equation, using a Krook-type model for the collision integral. Essentially all possible ion trajectories are considered and the regions of energy-angular momentum space are divided in the same fashion as was done by Bernstein and Rabinowitz. The moment method is essentially the Lees "two-stream Maxwellian" one. The idea is that a distribution function that satisfies the collisionless limit exactly is assumed, and collisional effects are accounted for in some average sense by taking sufficient moments of the Boltzmann equation to determine the various unknown parameters in the assumed distribution function. The analysis is rather too elaborate to reproduce here, although it is similar to the Bernstein–Rabinowitz calculation, except that additional terms arise because of the collision integral.

The cases calculated by CTW agree with Laframboise's results in the collisionless limit. It was found that as the ion-neutral Knudsen number λ_{in}/R is decreased, the ion current decreased relative to its collisionless value at the same values of ξ_p and χ_p, and the sheath region was found to extend further into the plasma. Enough cases were calculated to establish these trends, but not enough to cover a sufficient range of the parameters important to the experimentalist.

The theory of Bienkowski and Chang (1968) for a spherical probe, like the CTW theory, takes into account essentially all possible charged particle orbits within the framework of a moment method. However, "exact" results were obtained only for the asymptotic thin sheath limit. Although suggestions are given by Bienkowski and Chang for approximate procedures which could be used to relax these conditions, the range of conditions covered by the analysis is still somewhat limited.

Somewhat simpler approaches were used by Wasserstrom *et al.* (1965), and by Self and Shih (1968). Wasserstrom *et al.* used an approach similar to CTW, but considered only straight-line trajectories. Their results and the

CTW results agree in the limit of very small $|\chi_p|$, but the Wasserstrom *et al.* results become progressively less accurate as the probe potential is made more negative. However, their analysis is most applicable to the near-continuum limit, whereas the only results obtained by CTW were in the near-collisionless regime, so a comparison between the two approaches is not too meaningful. The approach of Self and Shih is essentially a modification of the ABR formulation for cold ions (radial motion), to account for collisions. The modification takes the form of a collisional friction term in the ion-momentum equation which for the collisionless case is Eq. (2.11) and which with collisions becomes for the ion velocity

$$v_r \frac{dv_r}{dr} = -\frac{e}{m_i}\frac{d\varphi}{dr} - \nu_{in} v_r \tag{2.20}$$

where ν_{in} is the ion-neutral collision frequency. This equation is solved together with the ion-continuity equation $I_i = 4\pi r^2 e N_i v_r$, and Poisson's equation, Eq. (2.13) with $v_s(-\chi^*)^{1/2}$ restored to v_r. A plot of their results is presented in the form of curves of a dimensionless current J (which in fact is equal to $\left.\frac{j_i^* \xi_p^2}{2\sqrt{\pi}}\right)$ plotted, for $\chi_p^* = -10$, versus ξ_p and the collision frequency ν_{in} normalized by the ion-plasma frequency, ω_{p_i}. The range covered is $1 \leq \xi_p \leq 100$, and $10^{-3} \leq \nu_{in}/\omega_{p_i} \leq 1$. Essentially, the same method of data reduction proposed by Sonin is described by these authors. Some experiments, unfortunately limited by the nature of the apparatus to a rather narrow range of ν_{in}/ω_{p_i}, are also reported which appear to confirm the predictions of the Self–Shih theory. In a subsequent paper Shih and Levi (1971) present an approximate form of the Self–Shih results, and extend the analysis to the case of a cylindrical probe. The approximation used restricts the applicability of the results to small collisional effects.

Although the calculations required in the Chou–Talbot–Willis analysis make this analysis difficult to use, approximate forms of the results have been developed. Talbot and Chou (1969) started with the general expression obtained by CTW for the spherical probe.

$$\begin{aligned} \frac{j_{i,\infty}}{j_i} &= 1 + \frac{R}{\lambda_{in}} j_{i,\infty} \int_0^1 e^{\chi} d\xi \\ &= 1 + \frac{R}{\lambda_{in}} j_{i,\infty} I_c \end{aligned} \tag{2.21}$$

where j_i is the normalized ion-current density, and $j_{i,\infty}$ is its value in the collisionless limit. They then constructed approximations for the integral I_c in Eq. (2.21), in the limits of collisionless and collision-dominated flows which we shall denote, respectively, by $I_{c,\infty}$ and $I_{c,0}$. Because $I_{c,\infty}$ and $I_{c,0}$

are of the same order of magnitude, they used a simple Knudsen number interpolation formula of the form

$$I_c = I_{c,\infty} + \frac{(I_{c,0} - I_{c,\infty})}{1 + \lambda_{in}/R} \tag{2.22}$$

to express I_c in the range between the two limits. With $j_{i,\infty}$ known from Laframboise's results as a function of χ_p, ξ_p, and ϵ, and I_c having been constructed as a function of these parameters, if the Knudsen number λ_{in}/R is known then the probe current j_i in the presence of collisions can be obtained. This approach was followed also for ion-current collection by a cylindrical probe by Talbot and Chou.

An even simpler approach was used by Schultz and Brown (1955), Sutton (1969), and later by Thornton (1971), who gave some additional justification for the procedure. In effect all of these investigators used a simple interpolation formula for the ion current itself (Sutton used this for the electron current) of the form

$$j_i = \frac{j_{i,\infty}}{1 + j_{i,\infty}/j_{i,0}} \tag{2.23}$$

Since $j_{i,0}$, the collision-dominated or diffusion-controlled limit of ion-current collection, is linear in λ_{in}, when $\lambda_{in} \to \infty$ then $j_i \to j_{i,\infty}$, and when $\lambda_{in} \to 0$, $j_i \to j_{i,0}$, and one can see that the approach is similar to the Knudsen number interpolation used by Talbot and Chou. From Eq. (2.23) it follows that the current collection to a probe of any geometry in the transition regime can be estimated if the collisionless and collision-dominated currents for that geometry are available. In this connection it is worth noting that better results for the collision-dominated limit for the cylinder are now available from the work of Inutake and Kuriki (1972) who applied the method of Su and Lam (1963) to the case of an ellipsoidal probe at highly negative potential.

The usefulness of these approximate transition-regime theories has been tested experimentally by Kirchhoff *et al.* (1971), Dunn and Lordi (1970), and Thornton (1971). Thornton found that the Talbot–Chou (1969), Self–Shih (1968) and his own results were all in quite good agreement with one another and represented his data well for both cylindrical and spherical probes. The Thornton method is probably the easiest to use. In the application of his method to a probe measurement in which the ion number density is initially unknown, it is of interest to point out that, just as in the case of the collisionless limit, the quantity $j_{i,0}\xi_p^2$ (or $j_{i,0}^*\xi_p^2$) is independent of N_∞. Hence one can construct transition-regime curves of $j_i\xi_p^2$ versus ξ_p for selected values of Knudsen number, probe-potential and ion-electron temperature rates, and follow the same approach described by Sonin to obtain N_∞ from a probe measurement in the transition regime without iteration.

The effect of electron-neutral collisions on electron-saturation current to cylindrical and spherical probes in a stationary plasma has been examined by Peterson (1971), using the Talbot–Chou approach and obtaining similar results. However, in the electron-retarding region of the probe characteristic, which is generally used to infer the electron temperature, the effect of electron-neutral collisions on the probe current is not well understood. There is evidence; both theoretical and experimental, that the classical $j_e = \exp\left(\frac{-e\varphi_p}{kT_e}\right)$ behavior is sufficiently altered by collisions that the method of obtaining T_e from the slope of a plot of $\ln j_e$-versus-probe potential φ_p no longer holds. Kirchhoff *et al.* (1971) have discussed the effect of electron-neutral collisions on the determination of electron temperature and have concluded on the basis of both theory and experiment that a double cylindrical probe is less sensitive than a single one to collisional effects, and hence that double probes may often be used to determine electron temperature under conditions in which single probes may give spurious results.

All of the aforementioned analyses for the effect of collisions on current collection by probes are concerned with *charged-neutral* collisions, which act to reduce the current below that of the collisionless limit (say, the Laframboise value). The effect of *ion–ion* collisions, under circumstances where ion-neutral collisions are negligible, appears to be just the reverse, at least for the case of the cylindrical probe. This is the conclusion reached as a result of a reappraisal of the measurements made by Sonin (1966) on cylindrical probes aligned with the flow direction under conditions where $\lambda_{in}/R \gg 1$ but $\lambda_{ii}/R < 1$. These data are reproduced in Fig. 2-6. Although a satisfactory independent

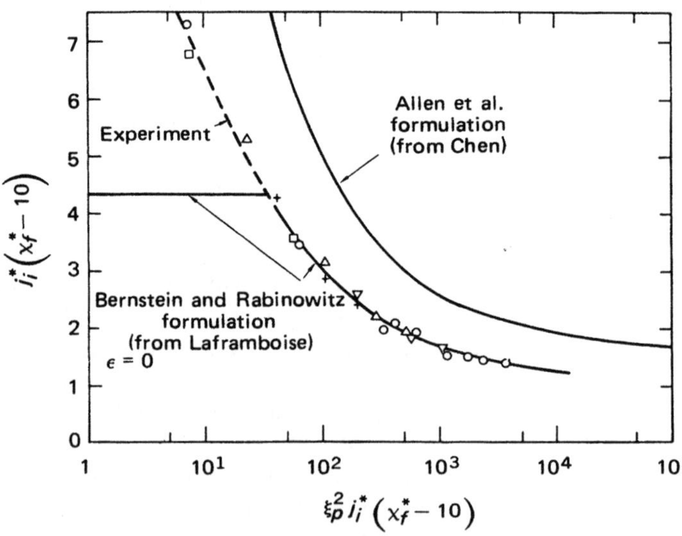

Fig. 2-6. Sonin's data for ion-current collection by an aligned cylindrical probe in a flowing plasma. (From Sonin, 1966.)

measurement of the free-stream ion density was not available to Sonin, he observed that his data followed the Laframboise prediction for the relative variation of $j_i^*\xi_p^2$ with j_i^* over their entire range, except when the OML was reached. The Laframboise calculation presented in this fashion predicts that when $j_i^*\xi_p^2$ is reduced to the value corresponding to the onset of orbital motion, the current j_i^* reaches a constant value and does not further increase with additional reduction in $j_i^*\xi_p^2$. It can be seen from Fig. 2.6 that Sonin found that j_i^* continued to increase as $j_i^*\xi_p^2$ decreased, and he attributed this to ion–ion collisional effects, having noted that λ_{ii}/R was not large in his experiments. This conjecture appears to be correct. However, a reassessment of these data by Hester and Sonin (1970) suggests that ion–ion collisional effects may be important not only in the orbital-motion regime, but over the entire range of $j_i^*\xi_p^2$, and that the data presented in Fig. 2-6 should be shifted upward to lie on the Chen (1965b) prediction for cylindrical probe current based on the Allen–Boyd–Reynolds model of radial motion. It was pointed out in Sec. 2.3 that whereas in the limit of $\epsilon = T_i/T_e = 0$, the Bernstein–Rabinowitz and Allen–Boyd–Reynolds analyses agree for the sphere, they do not for the cylinder; it has also been shown that for the cylinder the ABR radial motion model is not the correct limiting condition for $\epsilon = 0$, if the ion motion is rigorously collisionless. However, it is now believed that if λ_{ii}/R is not large, even though $\lambda_{in}/R \gg 1$ and ion–atom collisions are negligible, the effect of ion–ion collisions is to impede the tendency of ions to travel in curved trajectories and to produce essentially radial ion motion, thus giving rise to the greater ion current predicted by Chen based on the ABR model. As of the present, there is as yet insufficient experimental evidence to completely substantiate this hypothesis, and there is no satisfactory analysis that predicts the effect of ion–ion collisions in a self-consistent fashion, but the available evidence seems to favor the hypothesis.

2.6 Effect of Flow on Aligned Cylindrical Probes under Collisionless Conditions

As was remarked earlier, it seems plausible to assume that a cylindrical probe aligned with the flow direction will exhibit the same characteristics as predicted by Laframboise for cylindrical probe in a stationary collisionless plasma, if all the relevant Knudsen numbers $\lambda_{\alpha\beta}/R$ are $\gg 1$. This can also be assumed for the case in which only λ_{ii}/R does not satisfy the criterion, with characteristics predicted by the ABR model. This plausibility argument would seem to be substantiated by the experiments of Dunn and Lordi (1970) and by Graf and De Leeuw (1967). However, there is a situation for which the argument does not apply—when an "end effect" becomes important, and an additional parameter enters, the ratio l/λ_D of the probe length to the Debye length. This parameter may also be seen to arise from the introduction of the aspect ratio l/R, in the form of the product $(l/R)\,(R/\lambda_D)$.

The end effect was first discussed by Bettinger and Chen (1968) in connection with measurements made aboard Explorer 17. A sharp peak in the ion current was observed when the probe was precisely aligned with the flow. The same phenomenon was observed even more clearly in laboratory experiments carried out by Hester and Sonin (1969), as shown in Fig. 2-7.

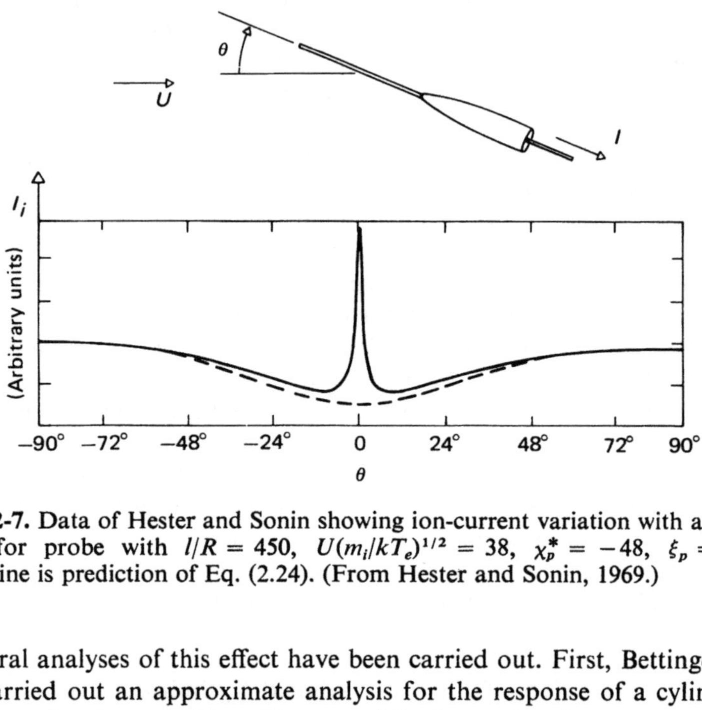

Fig. 2-7. Data of Hester and Sonin showing ion-current variation with angle of attack, for probe with $l/R = 450$, $U(m_i/kT_e)^{1/2} = 38$, $\chi_p^* = -48$, $\xi_p = 1/13$. Broken line is prediction of Eq. (2.24). (From Hester and Sonin, 1969.)

Several analyses of this effect have been carried out. First, Bettinger and Chen carried out an approximate analysis for the response of a cylindrical probe at angle of attack. Then Hester and Sonin produced a numerical solution for the aligned cylindrical probe. Sanmartin (1972) has produced an analytical study that parallels the numerical study of Hester and Sonin, and also Sanmartin (1972) has been able to carry out a more accurate analytical study of the angle-of-attack problem treated in approximate fashion by Bettinger and Chen.

Before discussing the results of these studies, particularly the results of Hester and Sonin, let us describe briefly in physical terms the nature of the end effect. It occurs for $\xi_p \ll 1$, that is, when the sheath radius becomes significantly greater than the probe radius and under conditions when, in a quiescent plasma, the probe would be operating in the OML regime. Now, Langmuir and Mott-Smith derived the approximate expression for current collection by a cylindrical probe at angle of attack θ to a flow with velocity U

$$I_{i,\infty} = 2eN_\infty URl\left[\sin^2\theta - \frac{e\varphi_p}{m_i U^2/2}\right]^{1/2} \tag{2.24}$$

and this is the expression shown by the dotted curve in Fig. 2-7. Observe that for $\theta = 0$, Eq. (2.24) reduces to

$$I_{i,\infty}(\theta = 0) = 2eN_\infty Rl\left[-\frac{2e\varphi_p}{m_i}\right]^{1/2} \tag{2.25}$$

which is identical to the OML Eq. (2.8a) in the limit $|e\varphi_p| \gg kT_i$. Equation (2.25) gives the current one would expect for an aligned cylindrical probe with $\xi_p \ll 1$, and does not account for the observed peak in the current. However, Eq. (2.25) is based on the model of current collection governed by an impact parameter b, as shown schematically for a probe in a transverse orientation in Fig. 2-8. But a probe in the aligned orientation, as shown also in Fig. 2-8, can collect particles not only through the lateral sheath surface area, but also through the end of the sheath. If the sheath radius is large enough, and if $U \gg (kT_e/m_i)^{1/2}$, then a significant number of ions can reach the probe with velocity U through the end of the sheath, as well as by motion across the cylindrical boundary of the sheath with a velocity, according to the Bohm condition, of the order of $\left(\dfrac{kT_e}{m_i}\right)^{1/2}$, and this in fact is the phenomenon responsible for the observed peak in ion current. Hence, the end effect should depend on the parameter

$$\tau_l = \frac{l}{\lambda_D}\frac{(kT_e/m_i)^{1/2}}{U} \tag{2.26}$$

because the relative importance of the current contribution from the end of

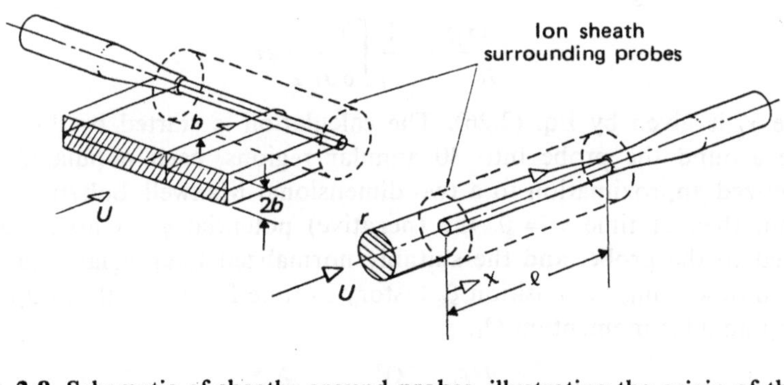

Fig. 2-8. Schematic of sheaths around probes, illustrating the origin of the end effect. (From Hester and Sonin, 1970.)

the sheath should depend upon the ratio of the product of the transverse velocity of the particles and the lateral surface of the sheath to the product of the directed flow velocity and the end area of the sheath. One may anticipate that for $\tau_l \gg 1$, the end effect will be negligible. Now, if the probe is not oriented exactly at $\theta = 0$, but turned at a slight angle to the flow direction, even though τ_l is not large many of the oncoming ions which enter the end of the sheath will have sufficient angular momentum to escape and will not be collected. This is the reason why the end effect, when it appears, is associated with a very small angular range around $\theta = 0$, and there is a rapid drop in current to the infinite probe value given by Eq. (2.24) for small θ. We can expect that the effect will also be a function of $\epsilon = T_i/T_e$, because the ion temperature is a measure of the random ion velocities in the lateral direction. In fact, it is this end effect, as was shown by Hester and Sonin, which is responsible for the apparent discrepancies between the Dunn and Lordi and Lederman *et al.* data and the Laframboise theory in the OML. When account is taken of the effects, then these data are found to agree with the theory throughout the entire range of ξ_p covered by the data.

Hester and Sonin approach the analysis of the end effect for aligned probes ($\theta = 0$) by formulating the problem as a one-dimensional unsteady one. They consider

$$j_i/j_{i,\infty} = f\left(\tau, \frac{\lambda_D}{R}, \frac{T_i}{T_e}, \frac{e\varphi_p}{kT_e}\right) \tag{2.27}$$

where $j_{i,\infty}$ is the current that would be collected by an infinitely long probe ($\tau_l \to \infty$), and

$$\tau = \frac{(kT_e/m_i)^{1/2} t}{\lambda_D} = \frac{x}{\lambda_D} \frac{(kT_e/m_i)^{1/2}}{U} \tag{2.28}$$

is a normalized time, which is related to the flow speed and the distance x from the leading edge of the probe in the steady-flow case. A mean current is defined according to

$$\frac{\bar{j}_i(\tau_l)}{j_{i,\infty}} = \frac{1}{\tau_l} \int_0^{\tau_l} \frac{j_i}{j_{i,\infty}}\, d\tau \tag{2.29}$$

where τ_l is given by Eq. (2.26). The calculation is started by dividing the space around the probe into 30 annular regions, each populated with a discretized approximation to a two-dimensional Maxwell–Boltzmann distribution, then at time $\tau = 0$, the (negative) potential φ_p is assumed to be applied to the probe, and the suitably normalized Lagrangian equation of motion describing the position ξ_k history in time for each kth group of ions having angular momentum Ω_k,

$$\frac{d^2\xi_k}{d\tau^2} + \frac{\Omega_k^2}{\xi_k^3} = -\frac{\partial \chi^*}{\partial \xi} \tag{2.30}$$

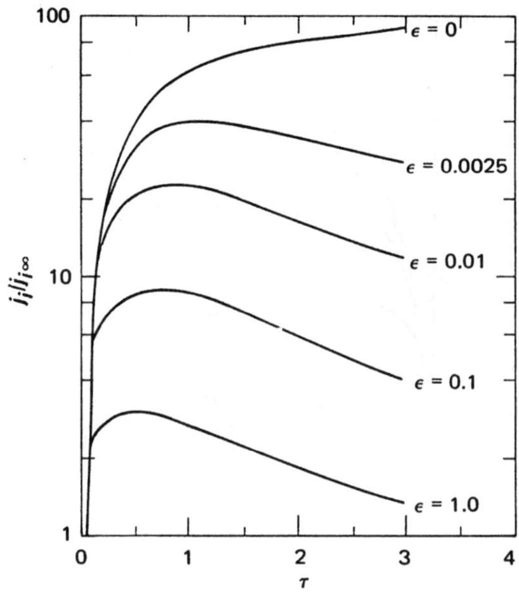

Fig. 2-9. Ion-current density to aligned cylindrical probe as a function of dimensionless time τ, for $\xi_p = 10^{-2}$ and $\chi_p^* = -15$. (From Hester and Sonin, 1970.)

together with Poisson's equation

$$\frac{1}{\xi}\frac{\partial}{\partial \xi}\left(\xi\frac{\partial \chi^*}{\partial \xi}\right) = e^{\chi^*} - N_k \tag{2.31}$$

are solved by successive approximations. The boundary conditions are

$$\chi^* = \chi_p^* \quad \text{at } \xi = \xi_p$$

$$\chi^* = 0, \qquad \left|\frac{\partial \chi^*}{\partial \xi}\right| \ll 1 \quad \text{at large } \xi \text{ (about } 5\lambda_D)$$

The current to the probe at time τ is obtained as the rate of accretion of charge, due to ions striking the probe.

Typical results of the Hester–Sonin calculation are shown in Figs. 2-9, 2-10, and 2-11. Figure 2-9 shows the instantaneous current density to a probe as a function of τ, for $\xi_p = 10^{-2}$, and for $\chi_p^* = -15$. The current starts at zero at $\tau = 0$, because of the way the problem was formulated, and rises rapidly to a maximum, around $\tau = 1$, for all finite values of $\epsilon = T_i/T_e$. Although the calculations were not carried beyond $\tau_l = 3$, the currents for all $\epsilon > 0$ decrease from their maxima as τ increases, and for sufficiently large τ, $j_i/j_{i,\infty}$ would approach unity. The exception to this is the result for $\epsilon = 0$,

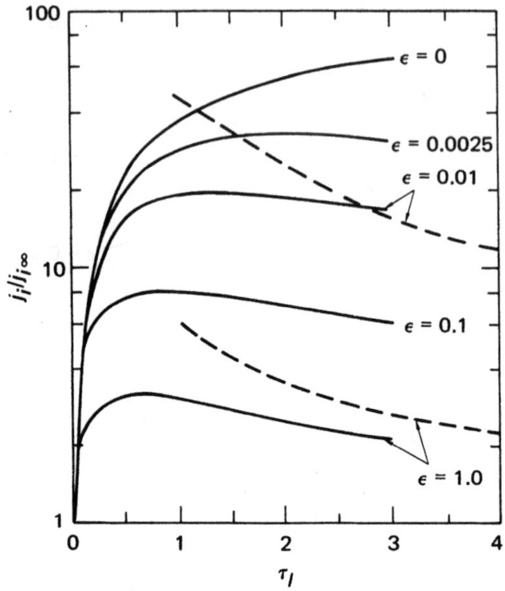

Fig. 2-10. Average ion-current density to aligned cylindrical probe as a function of the parameter τ_l, for $\xi_p = 10^{-2}$ and $\chi_p^* = -15$. Broken lines, theory of Bettinger and Chen. (From Hester and Sonin, 1970.)

which for large τ approaches a limiting value greater than unity. The reason for this, explained in more detail in the paper, is associated with the singular behavior of the cold-ion limit for cylindrical probes mentioned earlier in connection with the Allen–Boyd–Reynolds and Laframboise results.

Figures 2-10 and 2-11 show the average current to a probe, as a function of τ_l, for $\chi_p^* = -15$ and $\xi_p = 10^{-2}$ and 10^{-1}, respectively. Also shown for comparison are the results of Bettinger and Chen, which agree reasonably well in their region of overlap with those of Hester and Sonin. Several features of these results are worth noting. We see that the end effect is sensitive both to the ion temperature and to the Debye ratio ξ_p. For very small values of ξ_p, it is clear that physically very long probes are required to eliminate this end effect in a high-speed flow ($\tau_l \gtrsim 50$ for negligible end effect), and this fact must be considered in the design of experiments involving the use of cylindrical probes in flowing dilute plasmas. Another and perhaps even more intriguing conclusion that can be drawn is that pertaining to the transient response shown in Fig. 2-9. If one has a probe situation that under steady-state conditions corresponds to $\tau_l \gg 1$ (negligible end effect), then one can determine the ion density from Eq. (2.25) provided, of course, $\xi_p \lesssim 1$. However, if this same probe potential χ_p^* were suddenly applied at $\tau = 0$ and maintained at that value, the collected current would exhibit the response shown in Fig. 2-9. The overshoot in current above its ultimate steady-state value is a very strong function of ϵ, and thus the magnitude of the overshoot can be used to determine the ion temperature. As has been pointed out by Hester and Sonin, this is as far as is known the only direct practical method for measuring ion temperature. The numerical results for this very interesting transient probe response have now been supplemented by an analytical treatment due to Sanmartin (1972b), to which the reader is referred for further details.

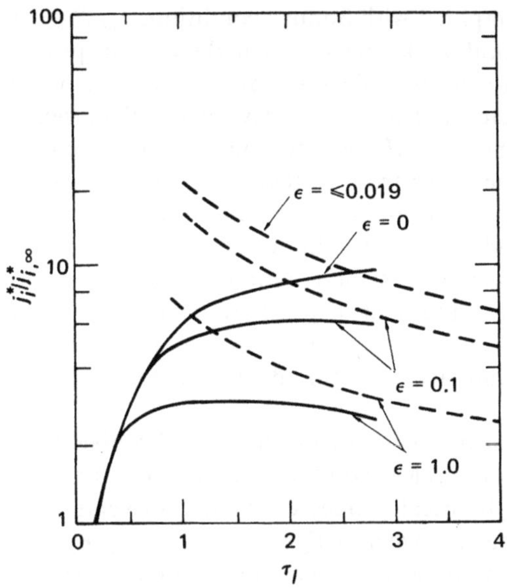

Fig. 2-11. Average ion-current density to aligned cylindrical probe as a function of the parameter τ_l, for $\xi_p = 10^{-1}$ and $\chi_p^* = -15$. Broken lines, theory of Bettinger and Chen. (From Hester and Sonin, 1970.)

The end effect we have been discussing is related to collisionless plasma flow past an aligned cylindrical probe. However, it appears that collisional effects may also produce phenomena similar to those observed under collisionless conditions for small values of τ_l, in particular a peak in ion current observed in the neighborhood of zero angle of incidence, rather than a monotonic variation with incidence of ion current such as is predicted by Eq. (2.24). This was, in fact, first noted by Sonin (1966), who found that, unlike the collisionless end effect, the peak was independent of probe aspect ratio l/R. Hester and Sonin (1970) later gave an explanation for this peak as being the result of ion–ion collisions; in effect, the probe when aligned with the flow will, because of ion–ion collisions, collect ion current more or less according to the ABR radial flow model, but when the probe is set at any significant incidence to the flow, the momentum of the flow will dominate over ion–ion collision effects and the current collected by the probe will be governed by orbital motion considerations. Jakubowski (1972) has reported ion-current measurements with cylindrical probes aligned with and at incidence to the flow, similar to the experiments of Sonin. Like Sonin, he also observed an ion current peak in the neighborhood of zero angle of incidence, which was independent of aspect ratio. The Knudsen numbers for Jakubowski's experiments lay in the ranges $0.4 < \lambda_{ii}/R < 1.5$, $10 < \lambda_{in}/R < 50$, which may be compared with Sonin's conditions $\lambda_{in}/R \gtrsim 10$, $0.03 < \lambda_{ii}/R < 0.5$. However, Jakubowski suggests that the current peak may be caused by "presheath" ion-neutral collisions rather than by ion–ion collisions as advanced by Hester and Sonin. In view of the Knudsen number ranges involved, the arguments of Hester and Sonin would appear the more plausible, although the issue is far from settled. It is perhaps worth mentioning that when λ_{in}/R decreases to values of the order of unity, or less, collisional effects on aligned cylindrical probes in a flowing plasma become much more complicated than in the case of a stationary plasma, as was found by Kirchhoff *et al.*, because the neutral gas flow field as well as the ion motion is modified by collisions; therefore Mach and Reynolds number effects also should be taken into account. No adequate theory exists for this transitional flow regime.

In some applications involving the use of probes in high-speed flowing plasmas under collisionless conditions at small values of ξ_p, it may be preferable to use cylindrical probes oriented transverse to the flow direction, particularly in the case where $kT_e/m_i \gg U^2 \gg kT_i/m_i$. For ion collection by such probes, the simplest result, valid in the infinite sheath limit $\xi_p \to 0$, is the one by Langmuir and Mott-Smith, Eq. (2.24) already cited. Various attempts have been made to obtain modifications to this result, which would account for the effects of finite sheath thickness, or, stated alternatively, reduction in the value of the impact parameter below its orbital value due to finite penetration of the probe's electric field. Among these attempts are the analyses of Clayden (1963), Smetana (1963), Kanal (1964), and Tan (1973). All of these analyses contain certain ad hoc assumptions that are introduced to deal with the very difficult problem posed by the nonsymmetric character of the potential distribution around the probe and the unknown (and also nonuniform) sheath thickness. Probably Smetana's results are as useful as any if the flow speed is not too high, and they have been used with reasonable success by several investigators (see Kang *et al.*, 1973).

However, there are several ways in which the behavior of a transverse cylindrical probe in a high-speed plasma flow differs from an aligned cylindrical probe, even neglecting such complications as end effects or collisional phenomena. In the limit $m_i U^2/kT_e \gg 1$, the ion current, according to Eq. (2.24), will be given simply (with $\theta = \pi/2$) by

$$I_i = 2eN_\infty URl \tag{2.32}$$

which is, in effect, what Clayden showed. However, the electron temperature cannot, in general, be determined directly from a plot of $\log j_e$-versus-probe potential (a value higher than the true electron temperature is obtained, which must be corrected for velocity effects), nor can the plasma potential be readily identified from a change in slope of the probe characteristic. And, finally, and perhaps most interestingly, electron-current saturation fails to occur under conditions where it would be present for a probe in a stationary plasma, because in essence the shielding effect of the probe electron sheath is destroyed by the high-speed ion flow. These phenomena have been demonstrated experimentally by Koopman (1971) and by Segall and Koopman (1973), and by Fournier (1971).

One of the more important applications of probes in flowing plasmas has to do with their use on sounding rockets and satellites for the measurement of charged particle densities at high altitudes. Many special probe configurations have been employed, and several theories have been developed which are applicable to the particular probe configurations and conditions associated with such measurements. Some examples are configurations involving probes with guard rings (Parker and Whipple, 1967), stagnation point probes (Sonin, 1967), and probes with screening grids (Whipple and Parker, 1969).

2.7 Summary

The following are some of the principal conclusions that can be drawn concerning the response of cylindrical and spherical electrostatic probes in the collisionless and transition regimes.

1. For cylindrical and spherical probes operated under completely collisionless conditions in a stationary plasma, the Laframboise (1966) theory gives an accurate description of the probe response, both for ion and electron currents, and can be used to obtain the electron density and temperature from the probe current-voltage characteristic curve. The monoenergetic ion model used by Bernstein and Rabinowitz (1959) is a very good approximation to the Maxwellian model used by Laframboise.

2. The cold-ion (or radial-motion) model of collisionless ion collection introduced by Allen *et al.* (1957) is a valid limiting condition for a spherical probe in stationary plasma, when $T_i/T_e \to 0$. The radial motion model is not a correct limiting condition for a cylindrical probe in a stationary plasma, when $T_i/T_e \to 0$, and the Bernstein–Rabinowitz results must be used instead.

However, the radial motion model appears to be a good representation for cylindrical probe ion-current collection when ion–ion collisions cannot be neglected.

3. The effect of ion-neutral collisions on ion-current collection by spherical and cylindrical probes in the transition regime in a stationary plasma can be estimated with satisfactory accuracy by means of a simple interpolation formula if the collisionless and collision-dominated limits of the probe ion current are known. This interpolation formula can thus be used to determine charged-particle density from the probe ion current in the transition regime. No such simple method exists for obtaining electron temperature from the electron-retarding portion of the characteristic curve of a single probe in the transition regime; however, double probes appear to be less susceptible than single probes to collisional effects in the transition regime and may often be used for electron-temperature determination in this regime when single probes give spurious results.

4. For flowing plasmas under collisionless conditions, cylindrical probes aligned with the flow direction give the same current-voltage characteristics as cylindrical probes in stationary plasmas in the collisionless regime, provided the end-effect parameter $\tau_l \gg 1$. For small values of τ_l, the ion current to an aligned cylindrical probe in a flowing plasma will exceed the orbital-motion-limited value because of additional ion flux reaching the probe through the end of the sheath. This end effect is limited to a quite small range of angle of incidence centered around the aligned position. Under conditions for which $\tau_l \gg 1$ and the end effect is negligible, ion–ion collisions can also produce a similar peak in ion-current collection by a cylindrical probe in a small range of incidence centered about the aligned position. This effect is independent of τ_l, and can thus be differentiated from the end effect by varying the probe aspect ratio l/R, which is porportional to τ_l.

5. Transitional effects on probes in flowing plasmas are more complex than for probes in stationary plasmas, and adequate methods for interpreting probe characteristics under flow conditions in the transitional regime are not available. A simple interpolation scheme such as has been found to be useful for probes in stationary plasmas fails under flow conditions for ion-neutral Knudsen numbers less than about unity.

6. Cylindrical probes oriented transverse to the flow direction and operated in the collisionless regime can be used to good effect in very high-speed flows. The interpretation of the ion-current collected by such probes is particularly simple. However, electron-current collection differs markedly from that found in stationary plasmas, in that the shielding effect of the electron sheath is obliterated by the high-speed ion flow, with the result that electron-current saturation does not occur. Electron temperatures can be obtained from the electron-retarding portion of the probe characteristic provided a velocity correction is applied.

References

Allen, J. E., Boyd, R. L. F., and Reynolds, P. (1957), *Proc. Phys. Soc. B.*, **70**, 297.
Bernstein, I. B. and Rabinowitz, I. (1959), *Phys. Fluids*, **2**, 112.
Bettinger, R. T. and Chen, A. A. (1968), *J. Geophys. Res.*, **73**, 2513.
Bienkowski, G. K. and Chang, K. W. (1968), *Phys. Fluids*, **11**, 784.
Bohm, D., Burhop, E. H. S., and Massey, H. S. W. (1949), *The Characteristics of Electrical Discharges in Magnetic Fields* (A. Guthrie and R. K. Wakerling, eds.), Chap. 2, McGraw-Hill, New York.
Chen, F. F. (1965a), "Electric Probes", in *Plasma Diagnostic Techniques* (R. H. Huddlestone and S. L. Leonard, eds.), Academic Press, New York.
Chen, F. F. (1965b), *Plasma Phys.* (*J. Nucl. Ener.* Part C), **7**, 47.
Chen, F. F., Etievant, C., and Mosher, D. (1968), *Phys. Fluids*, **11**, 811.
Chou, Y. S., Talbot, L., and Willis, D. R. (1966), *Phys. Fluids*, **9**, 2150.
Clayden, W. A. (1963), in *Rarefied Gas Dynamics*, 3rd Symposium (J. A. Laurmann, ed.), Vol. II, page 435, Academic Press, New York.
Dunn, M. G. and Lordi, J. A. (1970), *AIAA J.*, **8**, 1077.
Dunn, M. G. (1972), *AIAA J.*, **10**, 996.
Fournier, G. (1971), "Écoulement de Plasma sans Collisions autour d'un Cylindre en Vue d'Applications aux Sondes Ionospheriques," ONERA Publ. No. 137. *See also* Taillet, J., Brunet, A., and Fournier, G. (1973), "Behavior of a Positive Probe in High Speed Collision-Free Plasma Flow." in *Dynamics of Ionized Gases* (M. J. Lighthill, I. I. Imai, and H. Sato, eds.), University of Tokyo Press, Tokyo, Japan.
Graf, K. A. and De Leeuw, J. H. (1967), *J. Appl. Phys.* **38**, 4466.
Hall, L. S. and Fries, R. R. (1966), *Proc. 7th Int. Conf. Phenomena Ionized Gases*, **3**, 15–19.
Hester, S. D. and Sonin, A. A. (1969), in *Rarefied Gas Dynamics*, 6th Symposium (L. Trilling and H. Wachman, eds.), Vol. II, page 1659, Academic Press, New York.
Hester, S. D. and Sonin, A. A. (1970), *Phys. Fluids*, **13**, 1265.
Iachetta, F. A. and Smetana, F. O. (1969), in *Rarefied Gas Dynamics*, 6th Symposium (L. Trilling and H. Wachman, eds.), Vol. II, page 1783, Academic Press, New York.
Inutake, M. and Kuriki, K. (1972), "Characteristics of Cylindrical Langmuir Probe with the Effect of Collision," Book of Abstracts, 8th Rarefied Gas Dynamics Symposium, Stanford University, California. See also Inutake, M. (1968), Masters Thesis, Institute of Space and Aeronautical Sciences, Tokyo University, Tokyo, Japan.
Jakubowski, A. K. (1972), "The Behavior of Cylindrical Langmuir Probes in Hypersonic Flow of Nearly Collision-Free Plasma," Book of Abstracts, 8th Rarefied Gas Dynamics Symposium, Stanford University, California. *See also AIAA J.*, **10**, 988 (1972).
Johnson, E. O. and Malter, L. (1950), *Phys. Rev.*, **80**, 58.
Kanal, M. (1964), *J. Appl. Phys.*, **35**, 1687.
Kang, S. W., Jones, L. W., and Dunn, M. G. (1973), *AIAA J.*, **11**, 141.
Kiel, R. E. (1968), *AIAA J.*, **6**, 708.

Kiel, R. E. (1971), *AIAA J.*, **9,** 1380.
Kirchhoff, R. H., Peterson, E. W., and Talbot, L. (1971), *AIAA J.*, **9,** 1686.
Koopman, D. W. (1971), *Phys. Fluids*, **14,** 1707.
Laframboise, J. G. (1966), Univ of Toronto Institute of Aerospace Studies Report 100. Also in *Rarefied Gas Dynamics* (J. H. De Leeuw, ed.), Vol. II, page 22, Academic Press, New York.
Laframboise, J. G. and Parker, L. W. (1973), *Phys. Fluids*, **16,** 629.
Lam, S. H. (1965a), *Phys. Fluids*, **8,** 73.
Lam, S. H. (1965b), *Phys. Fluids*, **8,** 1002.
Langmuir, I. and Mott-Smith, H. M. (1926), *Phys. Rev.*, **28,** 727. Also in Suits, C. G., ed. (1961), *Collected Works of Irving Langmuir*, Vol. 4, pp. 99–132, Pergamon.
Lederman, S., Bloom, M. H., and Widhopf, G. F. (1968), *AIAA J.*, **6,** 2133.
Parker, L. W. and Whipple, E. C., Jr. (1967), *Ann. Phys* (U.S.), **44,** 126.
Peterson, E. W. and Talbot, L. (1970), *AIAA J.*, **8,** 2215.
Peterson, E. W. (1971). *AIAA J.*, **9,** 1404.
Sanmartin, J. R. (1972a), *Phys. Fluids*, **15,** 1134.
Sanmartin, J. R. (1972b), *Phys. Fluids*, **15,** 391.
Schultz, G. J. and Brown, S. C. (1955), *Phys. Rev.*, **98,** 1642.
Segall, S. B. and Koopman, D. W. (1973), *Phys. Fluids*, **16,** 1149.
Self, S. A. and Shih, C. H. (1968), *Phys. Fluids*, **11,** 1532.
Shih, C. H. and Levi, E. (1971), *AIAA J.*, **9,** 1673.
Smetana, F. O. (1963), "On the Current Collected by a Charged Circular Cylinder Immersed in a Two-Dimensional Rarefied Plasma Stream," in *Rarefied Gas Dynamics*, 3rd Symposium (J. A. Laurmann, ed.), Vol. II, page 65, Academic Press, New York.
Sonin, A. A. (1966), *AIAA J.*, **4,** 1588.
Sonin, A. A. (1967), *J. Geophys. Res.*, **72,** 4547.
Su, C. S. and Lam, S. H. (1963), *Phys. Fluids*, **6,** 1479.
Swift, J. D. and Schwar, M. J. R. (1971), *Electric Probes for Plasma Diagnostics*, Iliffe Books, London.
Talbot, L. and Chou, Y. S. (1969), in *Rarefied Gas Dynamics* (C. L. Brundin, ed.), Vol. II, page 1723, Academic Press, New York.
Tan, W. P. S. (1973), *J. Phys. D: Appl. Phys.*, **6,** 1188.
Thornton, J. A. (1971), *AIAA J.*, **9,** 342.
Wasserstrom, E., Su, C. H., Probstein, R. F. (1965), *Phys. Fluids*, **8,** 56.
Whipple, E. C., Jr. and Parker, L. W. (1969), *J. Geophys. Res.*, **74,** 2962.

CHAPTER III

Continuum Electric Probes

3.0 Introduction

The effects of collisions in the sheath formed around an electrostatic probe were discussed in some detail in Chapter II. In this chapter, we will study continuum electrostatic probes, which represent the limit of "many collisions" within the sheath. As outlined in Chapter I, one can identify two regimes of continuum probe operation and one hybrid case, depending on the relative magnitude of the smallest mean free path λ and the Debye length λ_D. They are

(a) $L \gg \lambda_D \gg \lambda$: collisional thin sheath
(b) $\lambda_D \gg L \gg \lambda$: collisional thick sheath
(c) $L \gg \lambda \gg \lambda_D$: collisionless thin sheath (hybrid dense case)

Here L is a characteristic length of the plasma region affected by the particular continuum probe, which will be defined more precisely in the course of subsequent discussions. When the probe consists of an entire body immersed in the plasma, which is the case we shall be mostly dealing with, then L is the characteristic dimension of the body. However, when the probe consists of an element embedded in the surface of a larger body, both the characteristic length of the body and the probe enter the problem. Also, the above thin-sheath criteria (a) and (c) are based on the premise that $|\chi_p| = O(1)$ at most. As will be discussed subsequently, χ_p must be included in more general definitions of thin and thick sheath criteria.

In terms of the domains discussed in Chapter I, the first two regimes (a) and (b) fall roughly in the region below the transitional sheath area demarcated in Fig. 1-2 of Chapter I. The last regime (c) is B-3. This regime, sometimes alluded to as the dense case, is the case where the sheath can be described by the free-fall considerations discussed in Chapter II, but the motion of carriers in the bulk of the plasma are determined from the continuum equations, (A.70), (A.71), (A.77), and (A.85) derived in the Appendix. The first two regimes are collision-dominated throughout the plasma and the motion of the carriers is determined by processes of convection, diffusion, and mobility, as described by the continuum equations.

In this chapter we will discuss the first two regimes in detail, and the third somewhat briefly. Where possible we will develop analytic expressions for probe current collection, display numerical solutions for probe characteristics, and make comparisons with experimental data where such data exist.

Our discussion is limited to weakly ionized plasmas, which will enable us to decouple the fluid mechanics from the electrical characteristics of the flow. The gas velocity, density, and temperature fields will therefore be assumed known and the quantities to be determined are charged-particle densities, the electric field in the plasma, and the electron temperature.

3.1 Some Physical Considerations

The relative importance of convection, diffusion, mobility, and charge generation on current collection are determined by nondimensional parameters that arise naturally, by nondimensionalizing Eqs. (A.70), (A.71), (A.77), and (A.85), as will be shown later in this chapter. To put the problem in perspective, however, we will discuss in broad terms some of these parameters, with the aid of the chart shown in Fig. 3-1.

As with collisionless and transitional electrostatic probes, there are two general plasma states encountered in continuum probe operations—the quiescent plasma, and the flowing plasma (either laminar or turbulent). These cases are demarcated by the magnitude of the convective terms in the governing charged-particle conservation equations, where the nondimensional

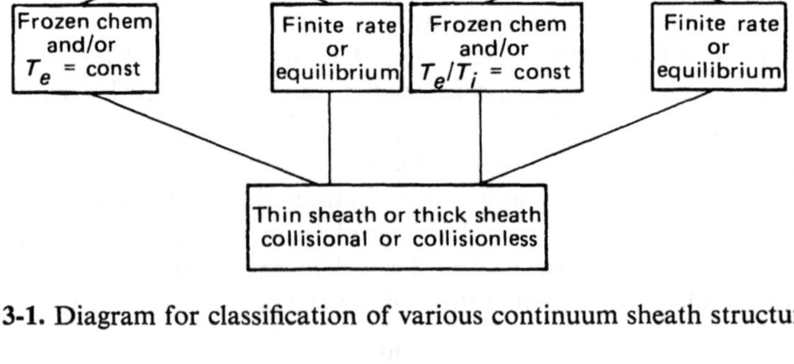

Fig. 3-1. Diagram for classification of various continuum sheath structures.

parameter is the electric Reynolds number, which is the product of the ion Schmidt number and the flow Reynolds number ($ReSc_i$).

Convection effects must be included when $ReSc_i$ is non-negligible. The usual boundary-layer-type approximations can be employed when $ReSc_i \gg 1$. All available analyses are for either the negligible or the large ($ReSc_i \gg 1$) convection limits.

For either the quiescent or large convective limits, the plasma may be incompressible, with constant properties, or it may be compressible, and have variable properties. Compressibility effects enter primarily through the parameter $\hat{\chi}_p = \chi_p(T_0/T_w)^n$, where $0 < n \leq 1$. Here χ_p is the nondimensional probe potential, and T_0 and T_w are reference and probe surface temperatures, respectively.

Whether compressible or incompressible, the plasma can have two limiting chemical or thermal states. These states are governed by the appropriate Damkohler numbers, which will be discussed later.

A brief explanation of the thermal states may be in order at this point.

In a thermal equilibrium state the collisional energy exchange between the electrons and the heavy-gas particles predominates over all other processes at each point in the plasma, and $T_e = T_i$ throughout. The electron-energy equation is then superfluous. In contrast, for plasmas in thermal nonequilibrium the electron temperature is governed by the electron-energy equation, which includes the effects of the electron energy-conduction and electric field as well as the effect of the collisional energy exchange between the electrons and heavy-gas particles. The limiting case of the thermally frozen plasma, in which the collisional energy exchange is negligible as compared to other processes, does not generally result in a constant T_e, particularly in the sheath region, even if the other properties such as T_i are constant. Therefore the electron-energy equation is necessary even in the case of the frozen thermal state. Often, however, much useful information pertaining to the effect of the nonequilibrium electron temperature on the electrical characteristics of the plasmas can be obtained by assuming the ratio T_e/T_i to be a constant, although possibly different from unity. Most of the constant-property analyses to be discussed here assume a constant T_e/T_i and neglect the electron-energy equation. In the following, we shall distinguish such analyses from the thermally frozen cases by referring to them as constant T_e/T_i analyses.

We come now to the probe-operation regimes noted in the Introduction, Section 3.0, which comprise the fourth level in Fig. 3-1. In discussing this level we should remember that the dichotomy of thin and thick sheaths is rather arbitrary. Many probe-operating conditions satisfy precisely neither the thin- nor the thick-sheath criteria set forth. However, this dichotomy aids us in the exploration of certain limiting properties of continuum probes. In terming a sheath "thin," it is usually implied that its effect on the rest of the plasma is negligible. The discussion of the sheath thicknesses will continue after the introduction of the nondimensional governing equations.

3.2 Governing Equations and Boundary Conditions

The governing equations that describe the continuum probe operation in weakly ionized plasmas consist of the species continuity equations, the electron energy equation, and the Poisson equation in addition to the overall continuity, momentum, and energy equations of the plasma. Because the degree of ionization is assumed very small, these overall equations are essentially those of the neutral gas ($T_n = T_i$) and to first order are unaffected by the presence of the ionized species. Following the derivations given in the Appendix [Eqs. (A.71), (A.70), (A.77), and (A.85)] we have for steady-state conditions.[4]

Ion Conservation

$$\rho\vec{u}\cdot\nabla C_i - \nabla\cdot\left[\rho D_i\left(\frac{C_i}{p_i}\nabla p_i + Z\frac{e}{kT_i}C_i\nabla\varphi\right)\right] = \dot{w}_i \tag{3.1}$$

Electron Conservation

$$\rho\vec{u}\cdot\nabla C_e - \nabla\cdot\left[\rho D_e\left(\frac{C_e}{p_e}\nabla p_e - \frac{e}{kT_e}C_e\nabla\varphi\right)\right] = \dot{w}_e \tag{3.2}$$

Electron Energy

$$\rho C_e\left[\vec{u} - D_e\left(\frac{1}{p_e}\nabla p_e - \frac{e}{kT_e}\nabla\varphi\right)\right]\cdot\nabla H_e = \nabla\cdot(K_{he}\nabla T_e) - \dot{w}_e h_e + \dot{w}_{he} \tag{3.3}$$

Electric Potential

$$\nabla^2\varphi = -\frac{e}{\sigma}\rho\left(Z\frac{C_i}{m_i} - \frac{C_e}{m_e}\right) \tag{3.4}$$

where the potential is used instead of the electric field intensity,

$$\nabla\varphi = -\vec{E}$$

In writing Eq. (3.1), the portion of the ion pressure diffusion represented

[4] Note change in notation from N to C for mass fraction, where

$$C_{i,e} = \frac{N_{i,e}m_{i,e}}{\rho}.$$

by the last term on the left-hand side of Eq. (A.71) has been considered negligible.

Boundary Conditions

These will be discussed more fully under specific applications. One generally specifies the following set of conditions.

At the fully catalytic and absorbing probe surface:[5]

$$C_{e,i} = 0$$

$$\varphi = \varphi_p \quad \text{(specified)} \tag{3.4a}$$

and

$$\frac{\rho C_e}{m_e} V_e \left(\frac{5}{2} kT_e - e\varphi\right) = K_{he} \frac{dT_e}{dy} \tag{3.4b}$$

Far from the probe

$$C_i = C_e \to C_0$$

$$\varphi \to 0$$

$$T_e \to T_{e0} \tag{3.4c}$$

Probe surface boundary conditions can be generalized to include non-catalytic or partially catalytic cases and electron- or ion-emission phenomena. More complete discussion on the above boundary conditions is given by Burke (1967). However, a brief discussion on Eq. (3.4b) is in order here.

The kinetic distribution of electrons in velocity space is always completely out of equilibrium within few mean free paths of a highly absorbent wall (see Chung, 1969). Therefore strictly speaking, the continuum concept is not applicable in such cases. If, however, the continuum equation (3.3) is to be employed all the way to the wall, then the complete kinetic nonequilibrium of the electrons mentioned above manifests itself as a singularity in the continuum equation at $y = 0$ as $K_{he} \to 0$. Chung (1965a) and Burke (1968) imposed the condition that $\partial^2 T_e/\partial y^2$ be finite at $y = 0$ on physical grounds. The electron-energy equation (3.3) then degenerates to Eq. (3.4b) at the wall. Since K_{he} approaches zero linearly with C_e, Eq. (3.4b)[6] gives a finite relationship between φ_w, T_{e_w}, and $\left(\frac{\partial T_e}{\partial y}\right)_w$. This expression essentially relates the

[5] The boundary condition $C_{e,i}=0$ is an acceptable continuum approximation because concentrations at the surface are very small compared to their values far from the probe. Detailed kinetic theory analysis shows that there is a lower bound to $C_{e,i}$ at the wall (cf. Persson, 1962).

[6] See Eqs. (A.53), (A.60), and (A.61).

electron-temperature drop across the last mean free path to the negative work done on the electrons by the electric field.

3.2.1 Nondimensional Equations

A set of the reference quantities is chosen as the characteristic length of the flow field L, and the various values at the reference point 0. A set of the dimensionless quantities are then defined in terms of the reference values: $\tilde{\vec{u}} = \vec{u}/u_0$, $\tilde{\rho} = \rho/\rho_0$, $\tilde{\nabla} = L\nabla$, $\tilde{T} = T/T_0$, $\tilde{C} = C/C_0$, etc.

Ion Conservation

$$(ReSc_i)\tilde{\rho}\tilde{\vec{u}} \cdot \tilde{\nabla}\tilde{C}_i - \tilde{\nabla} \cdot \left[\tilde{\rho}\tilde{D}_i\left(\frac{\tilde{C}_i}{\tilde{p}_i}\tilde{\nabla}\tilde{p}_i - Z\frac{\chi_p\tau}{\tilde{T}_i}\tilde{C}_i\tilde{\nabla}\psi\right)\right] = \mathscr{D}\tilde{\dot{w}}_i \tag{3.5}$$

Electron Conservation

$$(\beta ReSc_i)\tilde{\rho}\tilde{\vec{u}} \cdot \tilde{\nabla}\tilde{C}_e - \tilde{\nabla} \cdot \left[\tilde{\rho}\tilde{D}_e\left(\frac{\tilde{C}_e}{\tilde{p}_e}\tilde{\nabla}\tilde{p}_e + \frac{\chi_p}{\tilde{T}_e}\tilde{C}_e\tilde{\nabla}\psi\right)\right] = \beta\mathscr{D}\tilde{\dot{w}}_e \tag{3.6}$$

Electron Energy

$$(\beta ReSc_i)\,\tilde{\rho}\tilde{C}_e\tilde{\vec{u}} \cdot \tilde{\nabla}\tilde{H}_e - \tilde{\rho}\tilde{D}_e\tilde{C}_e\left(\frac{1}{\tilde{p}_e}\tilde{\nabla}p_e + \frac{\chi_p}{\tilde{T}_e}\tilde{\nabla}\psi\right)\cdot\nabla H_e$$

$$= \left(\frac{K_{he}T_e}{\rho C_e D_e H_e}\right)_0 \tilde{\nabla}(\tilde{K}_{he}\tilde{\nabla}\tilde{T}_e) - \mathscr{D}\beta\tilde{\dot{w}}_e\tilde{h}_e(h_e/H_e)_0 + \mathscr{D}_e\beta\tilde{\dot{w}}_{he} \tag{3.7}$$

Electric Potential

$$\chi_p\left(\frac{\lambda_D}{L}\right)^2\tilde{\nabla}^2\psi = \tilde{\rho}(\tilde{C}_i - \tilde{C}_e) \tag{3.8}$$

where Z is set to one and the nondimensional quantities are represented by tildes. The nondimensional parameters are defined as follows:

$$Re = \frac{\rho_0 L u_0}{\mu_0};\quad Sc_i = \left(\frac{\mu}{\rho D_i}\right)_0;\quad \beta = \frac{D_{i0}}{D_{e0}};\quad \tau = \frac{T_{e0}}{T_{i0}};\quad \psi = -\frac{\varphi}{\varphi_p};^{7}$$

$$\chi_p = \frac{e\varphi_p}{kT_{e0}};^{8}\quad \mathscr{D} = \frac{\dot{w}_{i0}L^2}{\rho_0 D_{i0}C_{i0}};\quad \mathscr{D}_e = \frac{\dot{w}_{he0}L^2}{\rho_0 D_{i0}C_{e0}H_{e0}};\quad \lambda_D = \sqrt{\frac{\sigma k T_{e0}}{e^2 N_{e0}}} \tag{3.8a}$$

In addition to these parameters, there are, of course, the governing para-

[7] The more customary use of nondimensionalization for φ is $\tilde{\psi} = e\varphi/kT_{e0}$. The rationale behind χ_p is to introduce a scaling factor, keeping ψ of order unity.

[8] Note that the present $\chi_p \equiv \chi_p^*$ of Chapter II.

meters of the neutral flow field such as the Mach number, the specific heat ratio, Prandtl number, etc. Of these, the ratio T_0/T_w (or equivalently ρ_w/ρ_0) enters explicitly in Eqs. (3.5)–(3.8) when carrying out compressibility transformations of the boundary-layer equations.

3.2.2 Definition of Parameters

Equations (3.5)–(3.8) in their general form present a difficult set of elliptic equations that cannot be solved analytically. Numerical solutions for the whole set are possible but not easy to obtain.

As discussed in Section 3.1, realistic probe-operating conditions often require simplified forms of Eqs. (3.5)–(3.8), which can be achieved by taking proper limits on one or more nondimensional parameters listed above. There is a certain arbitrariness in selecting these parameters but examination of Eqs. (3.5), (3.6), and (3.8) shows several to arise naturally. It is clear from Eqs. (3.5) and (3.6) that the first term on the left represents convective effects, the second term takes into account charge-carrier diffusion, and the third term represents charge transport caused by mobility. The term on the right is a source term that includes ionization–recombination effects in the flow. The same is true for the electron-energy equation [Eq. (3.7)] except the right-hand side contains electron-energy conduction and thermal equilibration terms in addition to the source term. Equation (3.8) relates the Laplacian of the potential to the electric-charge separation. The parameters or groups of parameters that should be considered are then

$ReSc_i = \dfrac{u_0 L}{D_{i0}}$ Ratio of convection to diffusion transport; sometimes called the electric Reynolds number

$\chi_{pT} = \dfrac{e\varphi_p}{kT_{i0}}$ Ratio of electric to thermal energy of ions

$\mathscr{D}_e, \mathscr{D}$ Damkohler numbers for thermal and chemical equilibration, respectively

$|\chi_p|\left(\dfrac{\lambda_D}{L}\right)^2$ Shielding length parameter based on electric potential

Another useful parameter employed first by Hammitt (1970) and Scharfman and Hammitt (1972) is the grouping $\dfrac{\chi_p}{Sc_i Re(\lambda_D/L)^2}$, which is essentially the ratio of mobility to convective transport.

The behavior of Eqs. (3.5)–(3.8) with respect to the convective parameter $ReSc_i$ is basically the same as that of neutral flows, which is well understood. The general role of the Damkohler numbers $\mathscr{D}$ and $\mathscr{D}_e$ in the governing equations are again the same as that of a Damkohler number in an electrically neutral, chemically reacting flow (see Chung, 1965b), i.e., they measure the degree of thermal and chemical nonequilibrium. However, an interesting point arises in the electron-energy equation, Eq. (3.7). The quantity $K_{he0}T_{e0}/(\rho_0 C_{e0} D_{e0} H_{e0})$ is of order Sc_i/Pr_n, which is usually of O(1), whereas the electron-energy source and convection terms are multiplied by $\beta \ll 1$. This is a manifestation of the fact that the electron thermal conductivity, K_{he}, is much larger than the heavy gas thermal conductivity. Therefore Eq. (3.7) shows that the influence of a change in the electron energy can extend through a region much greater than the species concentration and velocity boundary layers, and a much larger value of $\mathscr{D}_e$ is required to ensure equilibration of T_e with the heavy-gas temperature than the value of $\mathscr{D}$ required for equilibration of the chemical reaction of ionization–recombination in Eqs. (3.5) and (3.6).

It is noted here that D_e is also very much greater than D_i. However, the spatial distribution of electrons is closely tied to that of the ions through the electric field and, therefore, the extent of the electron concentration boundary layer is usually the same as that of the ions. In contrast, the electron thermal energy can be conducted to the fullest extent permitted by K_{he}. This basic behavior associated with the electron-energy equation was discussed by Chung and Mullen (1963), and by Burke (1968).

The remaining parameters, $\chi_p\tau$ and $|\chi_p|\left(\frac{\lambda_D}{L}\right)^2$, return us to the discussion of the sheath thicknesses initiated in Section 3.1. The behavior most peculiar to the flow of a continuum plasma is associated with these parameters.

The properties of Eqs. (3.5)–(3.8) to be discussed below are consistent with the similar discussion given in the Appendix, Section A.2.3, of the general properties of the Boltzmann and Poisson equations from which Eqs. (3.5)–(3.8) are derived for the continuum limit.

The basic properties of Eqs. (3.5)–(3.8) can be most conveniently studied from an examination of Eq. (3.8). First, in the limit $\left|\chi_p\left(\frac{\lambda_D}{L}\right)^2\right| \to \infty$ Eq. (3.8) degenerates to a Laplacian equation $\tilde{\nabla}^2\psi = 0$. Equations (3.5)–(3.7) are then decoupled from each other which implies that arbitrary charge separation can exist, unrestricted by space-charge effects. If we define, rather arbitrarily, the "sheath" as the region where charge separation is nonnegligible then the entire flow region is comprised of sheath in this case. This is then the limit of a thick sheath.

Next, let us consider the other limiting case, that of $|\chi_p|\left(\frac{\lambda_D}{L}\right)^2 \to 0$. We see from Eq. (3.8) that this limit is singular, which is evidenced by the disappearance of the highest-order derivative term in that equation. As is found in all singular perturbation problems, for a finite value of $|\chi_p|\left(\frac{\lambda_D}{L}\right)^2$, however small, there must exist an inner region[9] (sheath) within which the left-hand side of Eq. (3.8) must be retained (mathematically by an appropriate "stretching" of variables). In this sheath the right-hand side of Eq. (3.8) is also nonnegligible, which means that the charge separation is nonnegligible. In the "outer" region, where the left-hand side of Eq. (3.8) may be set equal to zero, charge separation is negligible, and therefore it can be identified as a quasineutral region in which there exist approximately equal numbers of positive and negative charges.

Obviously the stretching parameters and therefore the sheath thickness are functions of $|\chi_p|(\lambda_D/L)^2$, according to Eq. (3.8). For the plasmas in thermal equilibrium or with assumed constant $\frac{T_e}{T_i}$ the sheath thickness is mainly determined by the parameter $|\chi_p|(\lambda_D/L)^2$ provided that χ_p is of order one. When $|\chi_p| \gg 1$, the thickness becomes a function of χ_p as well as of $|\chi_p|(\lambda_D/L)^2$, since it appears in Eqs. (3.5) and (3.6). These problems, which will be discussed subsequently in more detail, have been analyzed by the method of matched asymptotic expansions as well as by other methods.

3.3 Specific Applications to Probe Theory

The ultimate goal for the present study is to construct probe characteristics for the various regimes outlined above and arrive at analytic and/or empirical expressions for predicting electron number densities and temperatures in the plasma under consideration. It follows from the above general considerations that current collection by a continuum probe of a given geometry depends on a large number of parameters. If we define a dimensionless probe current j,

$$j = \frac{I}{A_p N_{e0} Zev}$$

where v is an approximate characteristic speed and A_p is the probe surface area, it follows from Eqs. (3.5)–(3.8) that

$$j = j[(\lambda_D/L)^2, \chi_p, \chi_p\tau, ReSc_i, \mathscr{D}, \mathscr{D}_e, T_0/T_w] \tag{3.9}$$

Clearly, there is no simple way of obtaining an explicit, general expression for the current density in the above form, especially when one notes that geometry effects could play a significant role in the current-voltage (CV) characteristics of a probe as well. Historically, continuum probe theories have been developed by employing various limiting cases for one- or quasi-one-

[9] Note that here the term "inner region" is employed in a generic sense to mean the sheath. As will be discussed later, this inner region may be divided into several subregions in certain cases.

dimensional flows, displayed in Fig. 3-1. These theoretical and experimental developments will be discussed systematically subsequently, beginning with the quiescent case and then proceeding to the more complex cases demarcated in Fig. 3-1.

3.3.1 Small Convection Limit, $ReSc_i \to 0$

The simplest case of continuum probe operation is obtained in the one-dimensional, quiescent plasma limit. The earliest work in this limit was carried out for spherical and cylindrical probes by Boyd (1951) and Zakharova *et al.* (1960). An improved theoretical model for spherical probes was described by Su and Lam (1963) for thin sheaths with $\mathscr{D} \to 0$ and constant properties including constant T_e/T_i. In the limit $|\chi_p|\ (\lambda_{D0}/R)^2 \to 0$,[10] the exact sheath equations for spherical and elliptical probes were solved numerically by Cohen (1963, 1967, 1970). The Su–Lam and Cohen papers form the foundations of much further work on the subject. For example, the constant property analysis of Su and Lam (1963) has been extended to cylindrical geometries by Su and Kiel (1966), Toba and Sayano (1967), slender axisymmetric probes by Whitman and Chien (1971), ellipsoidal probes by Inutake and Kuriki (1972), and to thick-sheath studies on spheres and cylinders by Kiel (1969). Extension to the case of variable properties was done by Thomas (1969), and Chapkis and Baum (1971), to the nonisothermal electron-temperature case by Jou and Cheng (1971), and to the case of chemical nonequilibrium by Carrier and Fendell (1970) and McAssey and Yeh (1970). Further numerical work on the Su and Lam model was carried out by Cicerone and Bowhill (1967). Baum and Chapkis (1970) obtained exact numerical solutions for spherical probes for a wide range of bias potential and values of the ratio $|\chi_p|(\lambda_{D0}/R)^2$. Bush and Fendell (1970) gave a critical review of the work of Su and Lam and of Cohen, and obtained explicit results in three distinct bias potential domains for $\lambda_D/R \to 0$. The salient features of the above investigations are discussed in the following sections.

A. Frozen Chemistry, Constant Property Plasma

For the case of unbounded, quiescent, constant property frozen-chemistry plasma, Eqs. (3.5)–(3.8) reduce to

$$\tilde{\nabla} \cdot (\tilde{\nabla}\tilde{N}_i - \chi_p \tau \tilde{N}_i \tilde{\nabla}\psi) = 0 \tag{3.10}$$

$$\tilde{\nabla} \cdot (\tilde{\nabla}\tilde{N}_e + \chi_p \tilde{N}_e \tilde{\nabla}\psi) = 0 \tag{3.11}$$

$$\left(\frac{\lambda_D}{R}\right)^2 \chi_p \tilde{\nabla}^2 \psi = \tilde{N}_i - \tilde{N}_e \tag{3.12}$$

[10] For a quiescent plasma, the characteristic length L of the region affected by the probe is of the order of the probe size R. Therefore $L = R$ in the dimensionless parameters we have discussed.

where R is the probe radius and $\tilde{N}_{i,e} = N_{i,e}/N_{e0}$. Equation (3.7) is superfluous with the assumption of constant T_e/T_i. At the probe surface $\tilde{N}_i = \tilde{N}_e = 0$ (fully catalytic), and $\psi = -1$ and χ_p is specified. In the undisturbed plasma $\tilde{N}_i = \tilde{N}_e = 1$, $\psi \to 0$.

In general, no closed form solutions are possible for Eqs. (3.10)–(3.12) and they must be solved numerically for a given geometry. For small values of the parameter $|\chi_p|(\lambda_D/R)^2$, it is possible to divide the plasma into two broad regions as discussed in Section 3.2.2, a quasi-neutral region where the electric field is weak and $N_i = N_e$. and the sheath region, which sustains most of the field strength and where charge separation is significant. Solutions obtained in these two regions are then matched through a transitional region. For very large values of the parameter $|\chi_p|(\lambda_D/R)^2$ analytic solutions of Eqs. (3.10)–(3.12) are possible because $\nabla^2\psi = 0$, as explained in Section 3.2.2. However, for most cases of practical interest $|\chi_p|(\lambda_D/R)^2$ is small and $\nabla^2\psi \neq 0$.

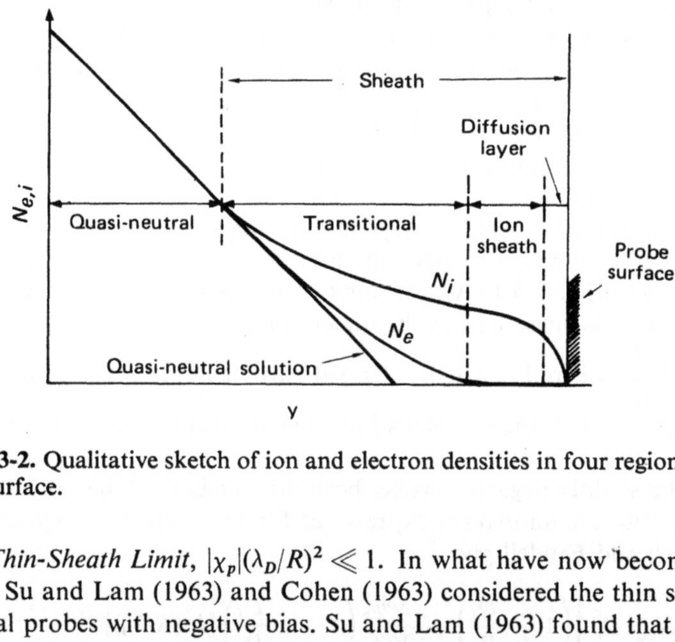

Fig. 3-2. Qualitative sketch of ion and electron densities in four regions near the probe surface.

1. *Thin-Sheath Limit*, $|\chi_p|(\lambda_D/R)^2 \ll 1$. In what have now become classic papers, Su and Lam (1963) and Cohen (1963) considered the thin sheath for spherical probes with negative bias. Su and Lam (1963) found that the solution for the highly negative probe is generally divided into four distinct regions in each of which different approximations are available. Without going into the mathematical details and quantitative estimates we will briefly discuss these four regions. As shown in Fig. 3-2. the plasma is almost neutral far away from the probe, and the solution is represented by the quasi-neutral solution $N_i \simeq N_e$. Because the potential is highly negative near the probe, the electron density is negligible compared with the ion density, and the solution is represented by the ion-sheath solution. A transitional region

exists between the quasi-neutral region and the ion-sheath region.[11] Immediately adjacent to the probe, an ion-diffusion layer is required to satisfy the wall-boundary conditions within the continuum description. The general behavior of the full solution is quite insensitive to the detailed structure of this diffusion layer. For a spherical probe with a highly negative sheath, Su and Lam have integrated Eqs. (3.10)–(3.12) numerically, assuming a Boltzmann electron-density distribution, viz.,

$$\tilde{N}_e = \exp(\psi) \tag{3.13}$$

In the limit $\lambda_D/R \ll 1$ and $|\chi_p| \to \infty$ they obtain the following analytic result for the current (in dimensional form) after neglecting the small electron contribution:

$$I = 4\pi R N_{e0} e D_i (-\chi_p \lambda_D/R)^{1/2}(1 + \tau) \tag{3.14}$$

Therefore, for a highly negative probe the ion current is a strong function of the temperature ratio τ and increases as the square root of the probe potential. In a later paper Lam (1968) obtained an expression very similar to Eq. (3.14) for the highly negative spherical probe in moderate pressures where the sheath is transitional. However, the new expression depends explicitly on the sheath radius R_s, which is not known a priori. Lam gives tabulated results for R_s as a function of χ_p and λ_D.

Cohen (1963) conducted an asymptotic analysis for moderately[12] biased probes and obtained extensive numerical results. The current-voltage characteristics obtained by Cohen for two values of τ are exhibited in Figs. 3-3(a) and (b). In these figures the dimensionless current density is given by $j_{i,e} = \dfrac{\Gamma_{i,e} R}{D_{i,e} N_{e0}}$, where $\Gamma_{i,e}$ are the charged-particle fluxes per unit area. In the limit of $\chi_p \ll -1$, Cohen's results join asymptotically with those of Su and Lam.

For the slightly negative probe, both Su and Lam (1963) and Bush and Fendell (1970) obtain explicit expressions for the current-voltage characteristics. Bush and Fendell give

$$I = 4\pi R N_{e0} e D_e \left\{\left(1 - \frac{D_i}{D_e}\right) - \frac{3e\varphi_p}{2kT_e}\left(1 + \tau\frac{D_i}{D_e}\right)[ln(R/\lambda_D)]^{-1} + \cdots\right\} \tag{3.15}$$

For a critical appraisal of Su and Lam (1963) and Cohen (1963), the reader is referred to Bush and Fendell (1970) who have attempted to reconstruct the

[11] A more useful analysis by Bush and Fendell (1970) actually uncovers two transitional regions, the first just interior to the quasi-neutral solution and the second, interior to and thicker than the first region.

[12] Bush and Fendell (1970) define more precisely the regime of applicability of Cohen's solution.

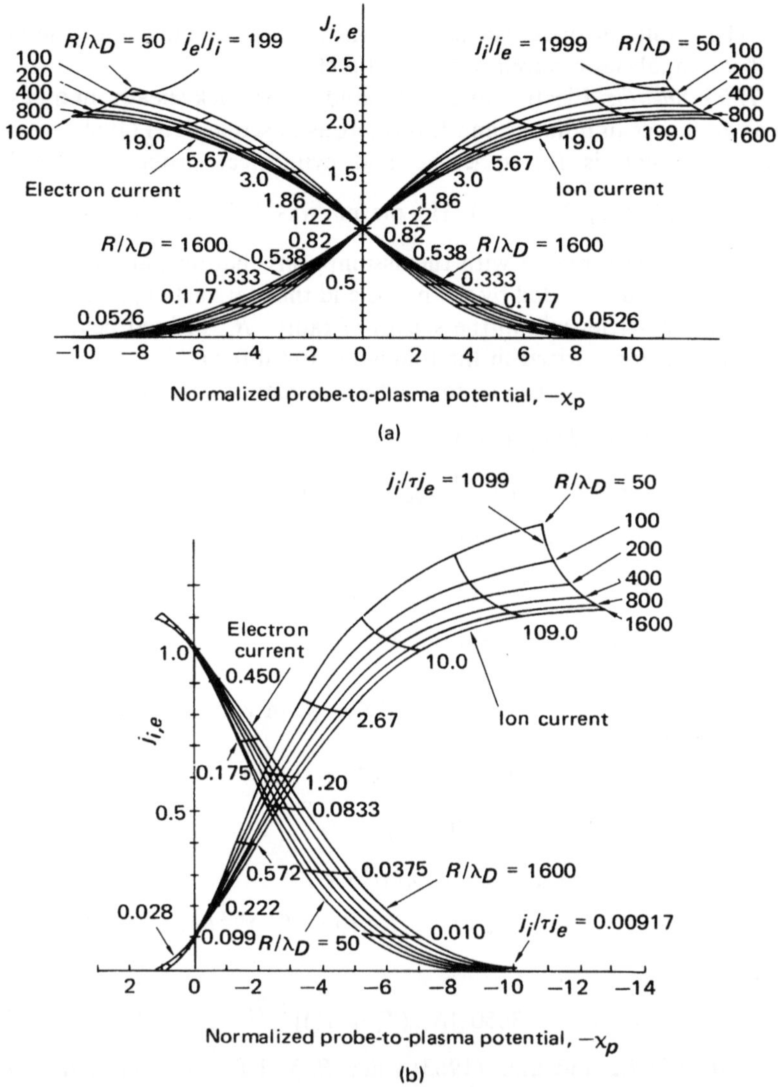

Fig. 3-3. (a) Normalized current characteristics for a spherical probe for various values of R/λ_D with $\tau = 1.00$ [from Cohen (1963)]. (b) Normalized current characteristics for a spherical probe for various values of R/λ_D with $\tau = 10$. (From Cohen, 1963.)

Su–Lam and Cohen analyses in terms of matched asymptotic expansions and have obtained refinements to their results.

Finally, Baum and Chapkis (1970) obtained exact solutions to Eqs. (3.10)–(3.12) by numerically integrating them for a sphere for all values of λ_D/R, with

$\tau = 1$. The results of these calculations will be discussed in conjunction with thick-sheath analyses covered in the next section.

2. *Thick Sheaths.* Electrostatic probe studies for thick sheaths in quiescent plasmas are fewer in number and less rigorous in scope. Kiel (1969) made an approximate analysis for spherical and cylindrical probes in the limit $|\chi_p|\left(\frac{\lambda_D}{R}\right)^2 \gg 1$, but $\lambda_D/R \leq O(1)$. The plasma was divided into a quasi-neutral and a simple sheath region with the assumptions that all particles attracted into the sheath are collected by the probe and these attracted particles are the only charged particles within the sheath of radius R_s (defined below). Kiel's results may be summarized in the following useful form.

For spheres, the electron or ion current in amperes is given by

$$I_{e,i} = 4\pi R R_s N_{e0}\mu_{e,i}k(T_i + T_e)$$
$$= 9.73 \times 10^{-15} Z\lambda_{en} R R_s N_{e0}(T_i + T_e)/(m_{e,i}T_{e,i})^{1/2} \quad (3.16)$$

where the ratio R_s of sheath radius to probe radius is given by

$$R_s = 1 + 388\{[N_{e0}(T_i + T_e)]^{-1/2}|\varphi_p|/R\}^{0.535} \quad (3.17)$$

and the units of the other quantities are

$$\mu_{e,i} = \frac{eD_{e,i}}{kT_{e,i}}, \quad N_{e0} \text{ in cm}^{-3}, \quad T \text{ in °K}, \quad \varphi_p \text{ in volts}, \quad R \text{ and } \lambda \text{ in cm},$$

$$m_{e,i} \text{ in grams}, \quad \text{and} \quad D_{e,i} \text{ in cm}^2\text{sec}^{-1}.$$

For a cylinder of length L_c (in cm)

$$I_{e,i} = 2\pi L_c N_{e0}\mu_{e,i}k(T_i + T_e)/\ln(\pi L_c/4RR_s)$$
$$= 4.87 \times 10^{-15} Z\lambda_{en} L_c N_{e0}(T_i + T_e)/[(m_{e,i}T_{e,i})^{1/2}\ln(\pi L_c/4RR_s)] \quad (3.18)$$

where

$$R_s = 1 + 3650\{[N_{e0}(T_i + T_e)]^{-1/2}|\varphi_p|/R\}^{0.62} \quad (3.19)$$

Note that as in Su and Lam (1963) when $R_s \gg 1$ for the sphere the current increases approximately as $\varphi_p^{0.5}$.

For the spherical probe, the results of Cohen (1963), Kiel (1969), and others can now be compared with the exact numerical solutions of Baum and Chapkis (1970) over a wide range of λ_D/R, with $\tau = 1$. Figure 3-4 shows the *CV* characteristics obtained by Baum and Chapkis for positive probe potentials for various values of R/λ_D.[13] These are by far the most useful results for spherical probes in unbounded, quiescent, constant property

[13] The characteristics for $\chi_p < 0$ can be obtained from Fig. 3.4 by letting $\chi_p \rightarrow -\chi_p$, $j_i \rightarrow j_e$, and $j_e \rightarrow j_i$.

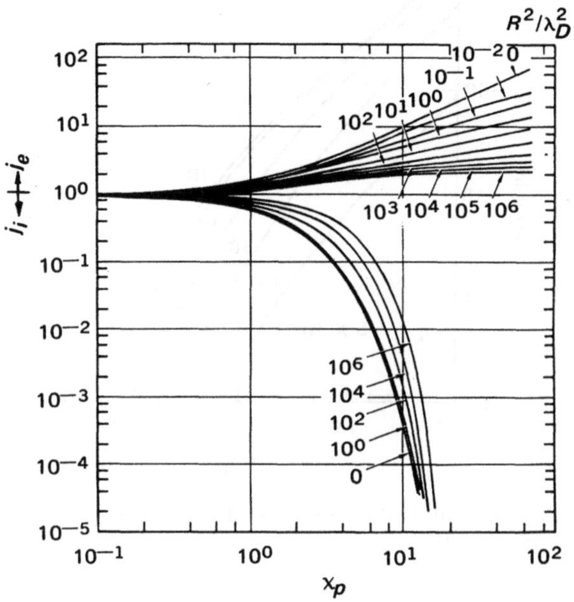

Fig. 3-4. Spherical probe characteristics for a quiescent plasma: Exact solution of Baum and Chapkis (1970).

plasmas. For large values of R/λ_D, the probe characteristics obtained by Cohen agree well with those of Fig. 3-4. However, because Cohen's analysis is an asymptotic one, valid for potentials of order unity and for thin sheaths, the agreement becomes poorer with increasing χ_p and decreasing R/λ_D. In fact the range of validity of Cohen's solution is shown in Fig. 3-5. In both figures

$$j_{e,i} = \frac{\Gamma_{e,i} R}{D_{e,i} N_{e0}} \tag{3.20}$$

and

$$\lambda_0 = [R\lambda_D^2/j_i(1 + j_e/j_i)]^{1/3} \sim \lambda_s \tag{3.21}$$

On comparing the Baum and Chapkis results with those of Kiel for a thick sheath, agreement is found within 20 per cent down to values of $R/\lambda_D =$ 1.

Interesting experimental verifications of the theories of Lam (1965, 1968) and Su and Lam (1963) have been obtained by Tanaka and Hirao (1972) with special probes operating in the electron-saturation limit where conditions varied continuously from a collisionless sheath limit into the continuum limit.

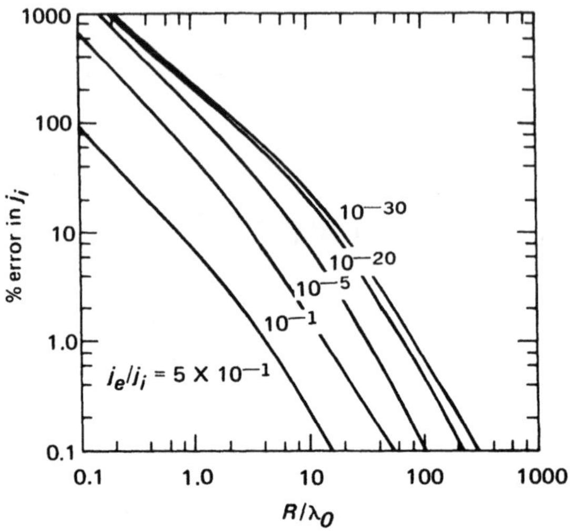

Fig. 3-5. Range of validity of Cohen's solution, according to exact solution of Baum and Chapkis (1970).

B. Frozen Chemistry, Variable Property Plasma

Until now we have been discussing probes in quiescent plasmas with constant properties. There are several investigations that deal with the variable property, quiescent, frozen plasma conditions. Thomas (1969) extended the Su and Lam analysis to the case of a spherical probe held at a much lower temperature than the surrounding gas, with an attendant nonuniform temperature field around the probe. Unfortunately he calculated the magnitude of this effect by using an incompressible form of the diffusion-flux expressions that does not completely account for the effect of gas density gradient on the probe current. He found little effect on the CV characteristics of the probe. Chapkis and Baum (1971) included the heavy-gas energy equation in the analysis of the cooled spherical probe. They neglected the electron diffusion caused by the gradient of T_e/T_i. The electron temperature was considered to be either frozen (T_e = constant) or in equilibrium with respect to the heavy-gas temperature. These analyses showed that the effect of the variable properties on the currents is most significant for small probe potentials. Chapkis and Baum (1971) found that the thin sheath ion saturation current for the equilibrium electron temperature is 23 percent smaller than that for the constant property plasma. For the frozen electron temperature, the ion-saturation current was found to be 6 percent larger than that for the constant property plasma, whereas the electron saturation current was 19 percent smaller than that for the constant property plasma.

Jou and Cheng (1971) included the electron energy equation in their analysis. Other properties, however, including the heavy-gas temperature, were considered to be constant. This analysis shows that the sheath scales are generally different from those given in the constant-property plasma theory.

The ion and electron currents approach their saturation values at lower absolute values of the probe potential than that predicted by the constant T_e theory, especially in the limit $\tau \to \infty$. An interesting result of Jou and Cheng's analysis is that it predicts ambient T_e values *higher* than those obtained from Cohen's constant-T_e analysis. When Little and Waymouth (1966) used Cohen's constant T_e/T_i theory to reduce their variable-T_e data, the electron temperature was predicted 40 percent too low. Therefore the variable-T_e analysis of Jou and Cheng show the correct trend, but the asymptotic nature of the analysis limits the usefulness of their results to highly negative probes as shown by the numerical solution of Barad and Cohen (1973).

C. Reacting Gas, Constant Property Plasma

We are now ready to look at finite-rate-chemistry effects on probe characteristics. The use of electrostatic probes in high density, weakly ionized gases, such as flames, requires an understanding of the effects of finite-rate ionization and recombination on the CV characteristics of the probe. If one assumes ionization and recombination to occur according to the following one-step process

$$A + B \underset{K_{rj}}{\overset{K_{fj}}{\rightleftarrows}} A^+ + e^- + B \tag{3.22}$$

where B is either an electron ($j = 1$) or a neutral ($j = 0$), and that the ionization and recombination rates K_f and K_r are functions of temperature, Eqs. (3.5)–(3.8) reduce to

$$\tilde{\nabla} \cdot [\tilde{D}_i(\tilde{\nabla}\tilde{N}_i - \chi_p\tau\tilde{N}_i\tilde{\nabla}\psi)] = \mathscr{D}(\tilde{N}_i\tilde{N}_e - S_j\tilde{N}_a)\tilde{N}_e \tag{3.23}$$

$$\tilde{\nabla} \cdot [\tilde{D}_e(\tilde{\nabla}\tilde{N}_e + \chi_p\tilde{N}_e\tilde{\nabla}\psi)] = \beta\mathscr{D}(\tilde{N}_i\tilde{N}_e - S_j\tilde{N}_a)\tilde{N}_e \tag{3.24}$$

$$\left(\frac{\lambda_D}{R}\right)^2 \chi_p\tilde{\nabla}^2\psi = \tilde{N}_i - \tilde{N}_e \tag{3.25}$$

where $\mathscr{D} = N_B^{1-j}R^2K_{rj}N_{e0}/D_i$, N_B is the number density of species B [Eq. (3.22)] and the neutral particle conservation equation is given by

$$\tilde{\nabla}^2\tilde{N}_a = \mathscr{D}(-\tilde{N}_i\tilde{N}_e + S_j\tilde{N}_a)\tilde{N}_e \tag{3.26}$$

where S_j, the degree of ionization at reference point 0, is given by $K_{fj}/K_{rj}N_{a0} = N_{e0}/N_{a0}$. It is instructive to note at this point that the Damkohler number can be represented as the ratio $\mathscr{D} = R^2/L_D^2$, thus introducing a new length scale $L_D = \left(\dfrac{\rho_0 D_i C_{i0}}{\dot{w}_{i0}}\right)^{1/2}$ in the problem. As in the case of electrically neutral chemically reacting flows, the chemical state is frozen for $\mathscr{D} \ll 1$ and is in equilibrium for $\mathscr{D} \gg 1$. In addition, the chemical reactions of ionization and recombination are limited to the ambipolar region when $\lambda_D/L_D \ll 1$, whereas the reactions within the sheath must be considered as well when λ_D/L_D is non-negligible.

There are few investigations that deal with electrostatic probes in reacting plasmas. The first detailed analysis for finite-rate gas-phase ionization and recombination was done by Cohen and Schweitzer (1968). A subsequent study was carried out by Carrier and Fendell (1970). Cohen and Schweitzer (1968) give an asymptotic analysis to first order in the Damkohler number to account for weak production effects in a continuum plasma over a spherical probe. Their results show that at negative probe potentials, the normalized ion current to the probe is increased as a result of charged particle production. In addition to asymptotic solutions for small and large Damkohler numbers, Carrier and Fendell solved Eqs. (3.23)–(3.25) numerically, both with neutrals and electrons serving as the third body of recombination. Their results show that the ion current collected at the probe increases approximately as the square root of the Damkohler number $\mathscr{D}$, especially when the probe potential χ_p becomes highly negative and the predominant reaction is ionization. This is in agreement with Cohen and Schweitzer (1968). The ion current decreases when the predominant reaction is recombination. Carrier and Fendell also show that the sheath structure is affected by gas-phase reactions only when the third body is a neutral.

3.3.2 Large Convection Limit, $ReSc_i \gg 1$

For $ReSc_i \gg 1$, a description of the plasma region affected by the probe requires at least two characteristic lengths—one in the streamwise direction and the other in the direction normal to the flow. The former is of the order R and the latter is of the order of the appropriate boundary-layer thickness δ. R is of the order of the nose radius in a blunt body flow, whereas it is of the order of the distance along the surface x for slender bodies.

In the various dimensionless parameters discussed $L = R$ except for the parameters $|\chi_p|\left(\frac{\lambda_D}{L}\right)^2$. For the latter $L = \delta$.

In the following discussions of the limit $ReSc_i \gg 1$, we shall employ the symbol L to represent R. Also, the reference point o is fixed along the boundary-layer edge unless otherwise specified. Subscripts o and δ are therefore employed interchangeably in the boundary layer discussions.

The pioneering work in this area was done by Lam (1964) who developed a general theory for the continuum flow of an incompressible weakly ionized gas about an arbitrary solid body with absorbing surfaces. His analysis was limited to the thin sheath case $|\chi_p|(\lambda_D/\delta)^2 \ll 1$. Subsequently, Lam (1965) and

Su (1965) extended Lam's work (1964) to compressible flows. More recently, Stahl and Su (1971) obtained detailed *CV* characteristics for incompressible flow over flush-mounted, highly negative probes where the sheath is one-dimensional, but thick.

Chung (1964) studied the electrical characteristics of compressible Couette and stagnation point boundary layer flows. The effect of non-equilibrium electron temperatures on stagnation point boundary layers was analyzed by Chung and Mullen (1963). The electrical characteristics of the viscous shock layer including both the effects of nonequilibrium electron temperatures and of chemical reaction were studied by Chung (1965a). Subsequently, Burke and Lam (1967) and Burke (1968) carried out an extensive analysis of the boundary layers in general, again including the various effects of nonequilibrium electron temperature and chemical reaction. Experimental results for flat-plate boundary layers were also obtained by Burke (1968) in these studies.

Chung and Blankenship (1966a) applied their boundary-layer analyses to the characteristics of double probes comprised of two parallel plates and to those comprised of two flush-mounted probes on pointed cones (Chung and Blankenship, 1966b). By extending these analyses and correlating the numerical results, Chung (1967) derived a simple diagnostic equation for the electron number density and temperature. Denison (1967) analyzed a problem similar to that of Chung (1966b) but for blunted cones, and obtained detailed *CV* characteristics in the limit $|\chi_p|(\lambda_D/\delta)^2 \ll 1$ and $ReSc_i \gg 1$.

Hoult (1965) considered the stagnation probe mounted on sounding rockets in the limit of large $|\chi_p|(\lambda_D/\delta)^2$, but with $ReSc_i \ll 1$. Sonin (1967) extended Hoult's analysis to $ReSc_i \gg 1$.

In all the above cases, the sheath is described by one-dimensional equations and, except for Chung (1965a), the convective effects in the sheath are neglected. deBoer and Johnson (1968) and Johnson and deBoer (1972) considered for a flat plate and a cylinder aligned with the flow the incompressible, large Reynolds number case with a one-dimensional sheath, where the sheath is thicker than the boundary layer and convective effects are important. Kulgein (1968) did a similar study for a cylinder in a cross-flow. Hammitt (1970) generalized the deBoer and Johnson results and applied them to cone and wedge probes.

The two-dimensional sheath was first discussed by Dukowicz (1969, 1970) who did a numerical study of the same problem for flat plates and cones using a two-dimensional Poisson equation in the limit of highly negative probe potentials. Baum and Denison (1971) solved the two-dimensional Poisson equation for a flush probe in an infinite conducting surface with infinitesimally thin insulators. The most complete analysis of multidimensional sheaths has been given by Russo (1972). A number of experiments have been conducted by several investigators that deal with thick sheaths in flowing plasmas. Some of these results will be discussed in the following sections.

Some of the aforementioned analyses are for incompressible flows; others take compressibility into account. For realistic flow situations, the incompressible, constant property case is of little interest. We will therefore be brief with the incompressible case and devote most of Section 3.3.2 to compressible, variable property plasmas.

A. Incompressible Constant Property Plasmas

Lam (1964) made a detailed study of the general behavior of the governing equations. For $ReSc_i \gg 1$, $|\chi_p| = O(1)$, and large probe surfaces of characteristic dimension L, Lam (1964) showed that the entire flow field naturally divides itself into three regions governing the electrical characteristics. This division is controlled by the parameters Re, Sc_i, T_e/T_i,[14] and λ_D/L. These three regions and their most important characteristics are discussed in the following.

Adjacent to the probe surface, there is a sheath the thickness of which is of the order $L[(ReSc_iT_i/T_e)^{-1/6}(\lambda_D/L)^{2/3}]$ wherein charge separation is non-negligible. In this region, the Poisson equation is retained along with the species conservation equations (see Section 3.2.2). Next there exists the ambi-

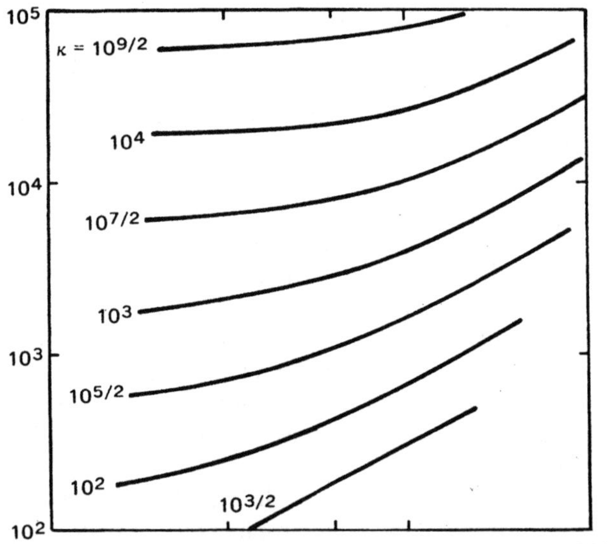

Fig. 3-6. Current-voltage characteristics for a flat plate probe in a flowing plasma with Blasius profile. $\kappa = (L^2/\lambda_D^2)/(Sc_i Re_x.)$ (From Stahl and Su, 1971.)

[14] Note that T_e/T_i is considered constant.

polar (quasi-neutral) region whose thickness is of the order $L[(ReSc_i T_i/T_e)^{-1/2}]$ wherein the charge separation is negligible and the Poisson equation is superfluous. In this region, the two-species conservation equations can be combined to a single species-conservation equation, which is identical in form to that of an electrically neutral species with an appropriate ambipolar Schmidt number (see also Chung, 1962, 1964). For Sc_i and T_e/T_i of order one, the order of thickness of this ambipolar region corresponds to that of the neutral gas-momentum boundary layer. Beyond the ambipolar region, there exists the inviscid potential region the thickness of which is of the order of $L[(ReSc_i T_e/T_i)^{1/2}]$. Through this region, the electric potential decays to zero. In this region, all nonuniformities are negligible except that of the electric potential. Hence the species conservation equations are trivially satisfied, and the Poisson equation becomes the Laplacian equation $\nabla^2\psi = 0$.

The ratio of the sheath to boundary-layer thicknesses can be readily formed from the above order relationships as

$$\frac{\lambda_s}{\delta} = \left(\frac{1}{Sc_i}\frac{T_e}{T_i}\right)^{1/6} Re^{1/3}\left(\frac{\lambda_D}{L}\right)^{2/3} \tag{3.27}$$

Now, we may define the "thin-sheath" limit, which we have defined thus far as the limit $|\chi_p|(\lambda_D/\delta)^2 \ll 1$, more precisely for constant-property boundary-layer flows as flows wherein

$$\left(\frac{1}{Sc_i}\frac{T_e}{T_i}\right)^{1/6}(Re^{1/2}\,\lambda_D/L)^{2/3} \ll 1 \tag{3.28}$$

when $|\chi_p|$ is of order one.

Note that Lam (1964) did not include χ_p in his scaling. Therefore χ_p has been tacitly assumed to be of order one, although cases for $|\chi_p| \gg 1$ were included in Lam's (1964) discussion.

Following Lam (1964) and Stahl and Su (1971) have recently analyzed the flat-plate probe in an incompressible flow employing a Blasius boundary-layer approximation. The results of the Stahl and Su analysis are quite useful in that they exhibit typical *CV* characteristics for moderate to highly negative probes in flowing plasmas with $T_e/T_i = 1$. We reproduce the Stahl and Su results in Fig. 3-6 where

$$j = \frac{4\pi L^2 I}{kT_i u_\delta\sqrt{ReSc_i}} \quad \text{and} \quad \kappa = \left(\frac{\lambda_D^2}{L^2}ReSc_i\right)^{-1} \simeq \delta^2/\lambda_D^2 \tag{3.29}$$

Figure 3-6 shows the continuously increasing ion current density j in the ion-saturation regime described in Chapter I. Large values of κ represent thin sheaths. With decreasing κ there is a progressively greater variation of j with χ_p. In addition, for each κ the j-versus-χ_p curves exhibit a relatively sudden change in slope at different values of the probe potential. This change in slope corresponds to the χ_p range where the sheath thickness λ_s becomes of the

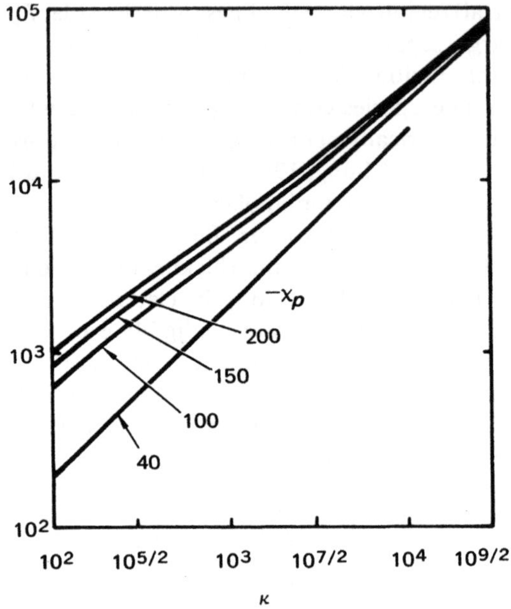

Fig. 3-7. Current collected by flat-plate probe in flowing plasma with Blasius profile, as function of parameter κ; $\kappa = (L^2/\lambda_D^2)\,/\,(Sc_i Re_x.)$ (From Stahl and Su, 1971.)

order of the boundary-layer thickness [it can be shown from Stahl and Su (1971) that $\lambda_s/\delta \propto \chi_p^{\frac{1}{2}}(\lambda_D/\delta)^{2/3}$]. A plot of j-versus-κ (Fig. 3-7) shows that $j \propto N_{e0}^p$, with $0.8 < p < 1.0$ depending on the probe bias χ_p.

It should be noted that the Stahl and Su sheath analysis is still one-dimensional, even when $\lambda_s = O(\delta)$. Recent experiments by Boyer and Touryan (1972), and two-dimensional sheath calculations by Russo and Touryan (1972) indicate that the simplified analysis of Stahl and Su is indeed a useful one that yields charged-particle number densities within a factor of two of the more accurate estimates (see discussion on thick sheaths in this section). A modest amount of numerical effort is required to apply the Stahl and Su analysis to a given specific case.

There have been other attempts, by deBoer and Johnson (1968) and by Johnson and deBoer (1972), to describe the current collection characteristics of flat plates in incompressible flows. In both of these the sheath is assumed to be *thicker* than the viscous boundary layer, the probe is at a high negative potential, and ion-diffusion effects are neglected. The first work is a theoretical study based on a quasi-one-dimensional representation of the ion flow within the continuum sheath. The second paper, which considers the two-dimensional sheath, gives an approximate theory both for flat plates and cylinders aligned with flow, based on the method of integral relations, which to lowest order yields the results of the quasi-one-dimensional solution.

David (1971) extended the thick-sheath flat-plate analysis of deBoer and Johnson to spherical probes and to cylindrical probes transverse to the flow direction. In the case of small-flow velocity, ion motion outside the sheath is governed both by convection and ambipolar diffusion. The solution for this region is matched to a symmetric sheath solution. For large-flow velocity, the ion motion outside the sheath is considered to be governed by convection only. The solution for the sheath in this case is determined by an iterative procedure, using Poisson's equation and the ion continuity equation.

Finally, Scharfman and Taylor (1971) used empirical relations to extend Kiel's approximate expression for ion-diffusion currents [Eqs. (3.17) and (3.18)] on spheres and cylinders to include convective effects. These results have limited usefulness but suggest types of corrections needed to apply results for quiescent plasmas to flowing ones.

B. Frozen Chemistry Compressible Variable-Property Plasmas

Electrostatic probes in compressible, nonisothermal flowing plasmas present a more realistic situation and are therefore more interesting than probes in incompressible flows.

As mentioned in the Introduction, the assumption of weakly ionized plasmas enables one to decouple the fluid mechanics from the electrical characteristics of the flow. In particular, it is assumed that the heavy particle temperatures $T_n = T_i$ are known from flow-field computations and that the electrons, because of their much smaller mass, do not appreciably affect $T_{n,i}$. The governing equations that define the probe behavior are then those given by (3.1)–(3.4). If we limit our compressible flow investigation to two-dimensional and axisymmetric boundary-layer flows, which represent the majority of practical cases encountered in probe operation, these equations become

$$\rho u \frac{\partial C_i}{\partial x} + \rho v \frac{\partial C_i}{\partial y} = \frac{\partial}{\partial y}\left[\rho D_i\left(\frac{\partial C_i}{\partial y} + \frac{Ze}{kT} C_i \frac{\partial \varphi}{\partial y}\right)\right] \tag{3.30}$$

$$\rho u \frac{\partial C_e}{\partial x} + \rho v \frac{\partial C_e}{\partial y} = \frac{\partial}{\partial y}\left\{\rho D_e\left[\frac{T}{T_e}\frac{\partial}{\partial y}\left(\frac{T_e C_e}{T}\right) - \frac{eC_e}{kT_e}\frac{\partial \varphi}{\partial y}\right]\right\} \tag{3.31}$$

$$\rho u C_e \frac{\partial H_e}{\partial x} + \rho v C_e \frac{\partial H_e}{\partial y} - \rho D_e\left[\frac{T}{T_e}\frac{\partial}{\partial y}\left(\frac{T_e C_e}{T}\right) - \frac{eC_e}{kT_e}\frac{\partial \varphi}{\partial y}\right]\frac{\partial H_e}{\partial y} = \frac{\partial}{\partial y}\left(K_{he}\frac{\partial T_e}{\partial y}\right) + \dot{w}_{he} \tag{3.32}$$

$$\frac{\partial^2 \varphi}{\partial y^2} = -\frac{e}{\sigma}\rho\left(Z\frac{C_i}{m_i} - \frac{C_e}{m_e}\right) \tag{3.33}$$

The boundary conditions are

at $y = 0$: $C_i = C_e = 0$, $\varphi = \varphi_p$,

and

$$\frac{\rho C_e}{m_e} V_e \left(\frac{5}{2} kT_e - e\varphi \right) = K_{he} \frac{dT_e}{\partial y} \tag{3.34}$$

for $y \to \infty$: $C_i = C_e = 1$, $\varphi = 0$,

and

$$T_e = T_{e\delta} \tag{3.35}$$

An order-of-magnitude estimate based on the compressibility transformation[15] of Chung (1965b) shows that the modification caused by compressibility of the effect of χ_p on probe electrical characteristics can be largely accounted for by replacing the term χ_p by $(\hat{\rho}/\rho_\delta)\chi_p$. [$\hat{\rho}$ is the boundary-layer-averaged density defined subsequently by Eq. (3.43).] One can approximate $\chi_p(\hat{\rho}/\rho_\delta)$ by $\chi_p(T_\delta/T_w)^n$ with $0 < n \leqq 1$. We note from this that in a compressible flow decreasing the wall temperature T_w is equivalent to increasing the surface potential χ_p, and vice versa.

As with the incompressible case, probe characteristics in compressible plasmas will be discussed in thin- and thick-sheath limits, respectively. We study the thin sheath first.

1. *Thin-Sheath Limit* $|\chi_p|(\lambda_D/L^2) \ll 1$. As we mentioned above the most complete work in this regime is that of Chung *et al.* (1962–1967), Burke and Lam (1967), and Burke (1968). Burke extended the general methods presented by Lam (1965), with special attention given to thermal nonequilibrium effects ($T_i \neq T_e$), and carried out detailed investigation of the role of various electron-energy transfer and exchange processes. In particular, Burke delineated four rather than three physically distinct regions which can differ in thickness by several orders of magnitude. The outer region, or the field fringing region, is the same as in Lam's incompressible analysis, and is coupled to the ambipolar and sheath regions via the conduction current term I_i. The electron-thermal layer is governed by the electron energy equation, and it is thicker than the viscous boundary layer because the effective Prandtl number Pr_e for the electron-energy equation is much less than unity (see also Chung and Mullen, 1963). The magnitudes of the changes in T_e in the various regions depend on whether the energy exchange between electrons and heavy particles is comparable to the electric field work in the outer regions or to thermal conduction in the viscous ambipolar region because here the electric field is weak. In the sheath the energy exchange must be compared with both thermal conduction and electric field work. For example, if we denote energy exchange

[15] Chung's (1965b) s–t transformation.

by ΔE, then if ΔE (conduction)/ΔE (collisional transfer) $\ll 1$, $T_e \simeq T_i$ and T_e can vary significantly in the boundary layer, as does T_i. If the above ratio is $\gg 1$, T_e will remain relatively constant throughout this region. The viscous ambipolar and sheath regions are also analogous to those of the incompressible case discussed by Lam. When the electrons in the sheath and ambipolar regions are assumed isothermal ($T_e = \text{const}$) the method of matched inner and outer expansion discussed in Section 3.3.1 is applicable, and indeed one recovers Cohen's sheath solution, modified by compressibility effects as discussed above.

For the thin-sheath case over a flat plate Burke (1968) arrived at the following expression for the net current j as a function of j_i/j_e, assuming T_e is known in the quasi-neutral region:

$$j = \frac{J}{eN_{e,\delta}u_\delta}\sqrt{Re_{\delta,x}} = \frac{1}{\sqrt{2}\,Sc_i}\tilde{J}_s\left(\frac{m_i}{m_e}\right)^{1/2}\frac{Q_{in}}{Q_{en}}\left(\frac{T_e}{T_i}\right)_\delta^{1/2} \tag{3.36}$$

where

$$\tilde{j}_s = -\beta_s T_s \mathcal{M}'(0)\left(\frac{T_\delta}{T_w}\right)\left(\frac{D_{e,s}}{D_{e,\delta}}\right)\left[\frac{(1 - j_e/j_i)(1 + T_i/T_e)_s}{1 + \beta_s \tau_s j_e/j_i}\right] \tag{3.37}$$

where $\mathcal{M}'(0)$, the normalized ion mass fraction gradient at the surface, is obtained from the solution of the ion-diffusion equation. Burke (1968) presented a few specific results of detailed numerical computations for Mach numbers 2.7, 2.9, and 6, which show the ion- and electron-saturation currents for the equilibrium and frozen electron temperatures.

One important consideration is that of relating the slope of the *CV* characteristic in the electron-retarding region to the free-stream electron temperature T_{e0}. In contrast to the familiar collisionless expression $\dfrac{d(\varphi)}{d(\ln j_e)} = \dfrac{kT_e}{e}$, Burke found the more complicated expression

$$\frac{d(\varphi)}{d(\ln j_e)} = \frac{kT_e}{e}\left[0.92 - A_1\frac{j_e}{j_i}\frac{T_{is}}{T_{es}} - A_2\frac{j_e}{j_i}\sqrt{Re_\infty}\right] \tag{3.38}$$

where A_1 and A_2 can be taken as constants in the range $10^{-2} \le j_e/j_i \le 10$. In general $A_2 = -\dfrac{\sqrt{2}\,(\beta_s\tau_s)}{1 + \beta\tau}\mathcal{M}'(0)N_{es}^2\dfrac{D_{es}}{D_{e\delta}}(1 + \tau)_s$ and A_1 is determined from the thin-sheath solution (Burke, 1967, page 129). It is clear from Eq. (3.38) that T_e is simply related to the *CV* slope *only if* A_1 and A_2 are small. In practice, the third term on the right is the largest, particularly for equilibrium plasma conditions. Because of the complexity of the problem, Burke did not propose any general diagnostic procedure based on his analysis. His results cannot be used directly to predict *CV* characteristics for the general case.

Talbot (1960) advanced the concept of matching the concentrations and gradients of the ionized species as well as the electric field intensity between the ambipolar and sheath solutions at the actual sheath edge. This concept was subsequently improved by Brundin (1964), resulting in a method for collisionless sheaths which utilizes $\mathscr{M}$ and $\mathscr{M}'$ at the actual sheath edge instead of necessarily at the wall. This method, although not as mathematically rigorous as that of Lam (1964), can be applied to all one-dimensional-sheath thicknesses smaller than the boundary layer thickness. Although Talbot studied only the collisionless sheaths, this study constitutes the first endeavor to analyze the electrical characteristics of a boundary layer.

Chung (1962, 1964) followed Talbot's concept in analyzing the thick as well as thin collision-dominated, one-dimensional sheaths. For Couette flow, a comparison of the results of the solutions obtained by matching the concentrations and gradients at the sheath edge with those obtained by the numerical integration of the complete flow field showed a very close agreement [within one or two percent for all sheath thicknesses (Chung, 1962)]. This matching method was also employed in the analysis of stagnation boundary layers with $T_e = T_i$. Although Chung (1964) considered stagnation boundary layers only, his solution is applicable to all similar boundary layers, with the appropriate definition of the similarity variable. It was shown, for $T_i = T_e$, that the sheath is sufficiently thin such that its effect on the ion-saturation current is negligible when the following inequality is satisfied

$$\hat{a} = \frac{2ls}{(\rho_\delta u_\delta)^2 r^{2\epsilon} \lambda_D^2} \sim \left(\frac{\delta}{\lambda_D}\right)_\delta \gtrsim 10^4 \tag{3.39}$$

where

$$s = \int_0^x \rho_\delta \mu_\delta u_\delta r^{2\epsilon} dx \qquad l = \left(\frac{\rho_w \mu_w}{\rho^* \mu^*}\right)^{0.2} \tag{3.40}$$

In Eq. (3.40) ρ^* and μ^* denote the values of these properties at the maximum temperature point in the boundary layer and subscript δ represents conditions at the boundary layer edge. For a low local Mach number M_δ we have $\rho^*\mu^* = \rho_\delta \mu_\delta$. The exponent ϵ is zero and unity, respectively, for two-dimensional and axisymmetric bodies, and r is the distance between the surface and the axis of symmetry. $\lambda_{D\delta}$ is the Debye length at δ, based on T_i. The parameter $\hat{a}$ represents a quantity of the order of the square of the ratio of the local density-modified boundary layer thickness to $\lambda_{D\delta}$ [$\hat{a}$ is closely related to the parameter κ defined in Eq. (3.29)].

For vanishingly thin sheaths satisfying the condition of Eq. (3.39), Chung's (1964) solution gives for the dimensionless ion-saturation current density, when $T_e/T_i = 1$,

$$j = \frac{J_{i,\text{sat}}}{eN_{e\delta}u_\delta} \frac{\sqrt{s/l}}{r^\epsilon \mu_\delta} = \frac{0.47}{\sqrt{2}} \left(\frac{2}{Sc_i}\right)^{2/3} \tag{3.41}$$

This is in fact the diffusion-controlled limit for the flux of any chemically frozen species (i) diffusing through the boundary layer to the wall. This implies, in the case for which the diffusing species are ions, that their motion is governed solely by the properties of the quasi-neutral region, and that the sheath plays no role in determining the species flux to the wall.

As was discussed in Chapter I, ion current generally increases continuously through the ion-saturation region. The onset of saturation, however, is readily recognizable. The ion current at the onset of saturation, such as the current corresponding to point *a* of Fig. 1-3, is usually employed in diagnostic work and this is the current referred to here as the "ion-saturation current". Note in Eq. (3.41) that explicit expressions for j for various self-similar flow configurations can be obtained by using the appropriate definitions of s obtained from Eqs. (3.40). Therefore we have as an example

$$j = \frac{J_{i,\text{sat}}}{eN_{e\delta}u_\delta}\sqrt{\frac{Re_x}{l}} \quad \text{for flat plate} \tag{3.42}$$

$$j = \frac{J_{i,\text{sat}}}{eN_{e\delta}u_\delta}\sqrt{\frac{Re_x}{3l}} \quad \text{for the cone} \tag{3.43}$$

and

$$j = \frac{J_{i,\text{sat}}}{2e(1+\epsilon)N_{e\delta}}\sqrt{\frac{\rho_\delta}{\mu_\delta\left(\dfrac{d\mu}{dx}\right)_\delta l}} \quad \text{for the highly cooled stagnation region} \tag{3.44}$$

where

$$Re_x = \frac{\rho_\delta u_\delta x}{\mu_\delta}$$

Chung's (1964) variable-property thin-sheath criterion, Eq. (3.39), can be compared to the Lam (1964) constant-property thin-sheath criterion, Eq. (3.28), for T_e/T_i and Sc_i of order unity. In Chung's (1964) work,

$$\left(\frac{\lambda_{D\delta}}{\delta}\right)^2 = \mathrm{O}\left[\left(\frac{\hat{\rho}}{\rho_\delta}\right)^2 \frac{1}{\hat{a}}\right] \tag{3.45}$$

where

$$\hat{\rho} = \delta^{-1}\int_0^\delta \rho\, dy$$

Since δ is of the order $LRe^{-1/2}$ in Lam's (1964) work, on substituting Eq. (3.45) for $(Re^{1/2}\,\lambda_D/L)^2$ into Eq. (3.28), Lam's criterion becomes

$$\left(\frac{\hat{\rho}}{\rho_\delta}\right)^{2/3} \hat{a}^{-1/3} \ll 1 \tag{3.46}$$

A comparison between Eqs. (3.39) and (3.45) shows that the Chung (1964) and Lam (1964) thin-sheath criteria are comparable up to $\hat{\rho}/\rho_\delta$ of order 10.

Chung and Blankenship (1966b), using the sheath solutions of Chung (1964), studied the behavior of flush probes located on sharp cones for high local-Mach-number flows where the ionization takes place within the boundary layer. They assumed that the ionization reaction was confined to a narrow region in the boundary layer in the neighborhood of its temperature maximum about the temperature peak and that the location of the maximum ionized species concentration, η_m,[16] corresponded to that of the temperature maximum. The expression derived for the maximum electron number density $N_{em} = N_e(\eta_m)$, for $T_e = T_i$, is

$$N_{em} = 1.9 \times 10^{18} Sc_i (Re_x/l)^{1/2} \frac{1}{u_\delta}\left(\frac{\rho_m}{\rho_\delta}\right) \eta_m J_{i,\text{sat}} \tag{3.47}$$

where l is given by Eq. (3.40) and N_{em} and u_δ are in particles per cm^3 and cm/sec, respectively. $J_{i,\text{sat}}$ is the ion-saturation current in amps per cm^2.

For Prandtl number of unity, the location of the temperature peak is given by the implicit relationship

$$f'(\eta_m) = \frac{H_\delta}{u_\delta^2}\left(1 - \frac{H_w}{H_\delta}\right) \tag{3.48}$$

where $f'(= u/u_\delta)$ is the derivative of the regular Blasius function the tabulation of which is found in most fluid mechanics textbooks, and H is the total stagnation enthalpy.

It is noted here that Eq. (3.47) is a quite general result in that it simply gives the electron-number density at η_m provided that η_m is the position of the peak electron-mass fraction. For instance, if one sets $\eta_m = \eta_\delta$, Eq. (3.47) gives $N_{e\delta}$, which is within a few percent of that according to Eq. (3.41) for a cone with the maximum electron mass fraction at the boundary layer edge.

Denison (1967) carried out a more detailed analysis of a closely related problem, that of a slightly blunted cone. In his analysis, he assumed that the ionization took place across the bow shock at the stagnation region of the blunted cone, and that the ionized species then diffused toward the wall as they were convected around the nose and along the cone in a chemically frozen state. For a slightly blunted cone, the shock layer gas is rapidly swallowed into the boundary layer, and therefore over the cone surface, except very close to the nose, the location of the maximum of the ionized species concentration can be expected to be inside the momentum boundary layer. The solution for this problem obtained by Denison (1967) gives for the peak electron density, under the assumption that $T_i = T_e$

[16] $\eta = \frac{r^\epsilon u_\delta}{\sqrt{2s}\,l}\int_0^y \rho\,dy$, which for a cone is $\eta = \sqrt{\frac{3u_\delta}{2l\rho_\delta u_\delta x}}\int_0^y \rho\,dy$

$$N_{em} = 3.68 \times 10^{18} Sc_i (Re_x/l)^{1/2} \frac{1}{u_\delta}\left(\frac{\rho_m}{\rho_\delta}\right) J_{i,\text{sat}} \tag{3.49}$$

where the units of the quantities are the same as those given for Eq. (3.47). The location of the peak electron density was found to be $\eta_{\max} \simeq 2$.

The accuracy of Chung and Blankenship's (1966b) approximate result, Eq. (3.47), can be readily checked against the more exact solution of Denison, Eq. (3.49). As has been stated, the maximum electron concentration exists at $\eta_m \simeq 2$, when the charged particles are generated at the nosetip only. Recall that η_m in the Chung and Blankenship analysis is simply the position of the maximum ionized-species concentration. Substitution of $\eta_m = 2$ into Eq. (3.47) gives a value of N_{em} for the slightly blunted cone, which agrees within about 3 percent with that obtained from Eq. (3.49).

On the other hand, Eq. (3.48) gives $\eta_m \approx 1$ for the position of maximum electron concentration in the cone boundary layer when the ionization source is boundary-layer viscous dissipation. With $\eta_m = 1$, Eq. (3.47) shows that electrons generated by viscous dissipation would produce a current $J_{i,\text{sat}}$, which is twice as large as the current resulting from ionization produced at the blunted nosetip, the case for which Eq. (3.49) was derived.

Chung and Blankenship (1966a) and Chung (1967) extended the method of Chung (1964) described earlier to the cases of frozen electron-neutral gas temperature equilibration (thermally frozen) and frozen chemical reaction, for the purpose of obtaining simple diagnostic relationships. As was shown in the detailed solutions of Chung (1965) and Burke and Lam (1967), the electron temperature varies little through the ambipolar region. Hence the ratio T_e/T_i is very high near the wall when the wall is strongly cooled in a high-energy flow. Because of the large value of T_e/T_i near the wall, Eq. (3.39) is no longer a valid criterion for the existence of a "negligibly thin" sheath. In fact the sheath can be of the order of the boundary-layer thickness for $a \sim 10^4$ if $(T_e/T_i)_\delta$ is sufficiently large.[17] What is more important than the actual thickness of the sheath is that the sheath plays a much greater role in determining the ion-saturation current when the plasma is thermally frozen than when it is in equilibrium, even when the physical thickness of the sheath is negligible. For instance, with $C_{is}/C_{i\delta} \approx 10^{-2}$ the effect of a physically thin sheath on the ion-saturation current is indeed negligible when $T_e = T_i$. On the other hand, such a sheath in a thermally frozen flow with a highly cooled wall has a large effect on the determination of the ion-saturation current and potential. For instance, Chung and Blankenship (1966a) found that the effect of the sheath on the saturation current was of first order up to $\hat{a} \approx 10^9$ if $(T_e/T_i)_\delta$ is sufficiently large.

In an application of the matching method of Chung (1964) for a thermally

[17] In Chung's parameter $\hat{a}$, λ_D is based on $T_{i\delta}$. Moreover, χ_p is based on $T_{i\delta}$.

frozen plasma, Chung and Blankenship (1966a) and Chung (1967) analyzed locally similar boundary layer flows using an approach wherein the quasi-neutral region was solved for approximately, whereas the sheath equations were integrated exactly, including the electron energy equation. A series of numerical solutions was obtained for highly cooled, low local Mach number ($M_\delta < 2$) flows for various values of Pr, Sc_i, and $(T_e/T_i)_\delta$. The range of $\hat{a}$ covered in the computation was between 10^4 and 10^9. The following expressions for dimensionless ion-saturation current density for thermally and chemically frozen, locally similar boundary layers was found to correlate the computed solutions to within about 20 percent.

$$j = \frac{J_{i,\mathrm{sat}}}{eN_{e\delta}u_\delta}\frac{\sqrt{s/l}}{r^\epsilon \mu_\delta} = \left[1 + \left(\frac{T_e}{T}\right)_\delta\right]\frac{(0.293 + 0.181\ Sc_i)}{\sqrt{2}\ Sc_i} \tag{3.50}$$

The relationships, Eqs. (3.42)–(3.44) apply here also.

Based on an analysis of their solutions, Chung *et al.* (1966a, 1967) also proposed that the CV relationship for a double probe could be employed to determine the electron temperature of the inviscid plasma through use of the formula

$$\chi_{p,s} = 2.1\left(\frac{T_e}{T}\right)_\delta \tag{3.51}$$

where $\chi_{p,s}$ is $|e(\varphi_{p,a} - \varphi_{p,b})/(kT_{i\delta})|$ shown in Fig. 1-3. Once $(T_e/T_i)_\delta$ is determined from Eq. (3.51), Eq. (3.50) can be employed to determine $N_{e\delta}$. Equations (3.50) and (3.51) were originally developed for the double-probe comprised of two parallel plates. However, the ion-saturation characteristics of double and single probes of the same geometry are identical provided that the maximum ion concentration is at the boundary-layer edge of both types of probes. Hence Eqs. (3.41), (3.50) and (3.51) hold for all locally similar, highly cooled boundary layers on single or double probes.

Scharfman and Bredfeldt (1967, 1970) have performed extensive measurements on flush-mounted probes with continuum flow conditions. The experiments were conducted in a shock tube in the range of $10^9 < N_{e\delta} < 10^{14}$ per cm^3 and -90 volts $< \varphi_p < -3$ volts. The experiments conducted at the pressure of 0.1 torr were found to agree very closely with Eq. (3.50) when $\varphi_p = -3$ volts, which corresponds to $e\,|\varphi_p|/(kT_{e\delta})$ of 14. For φ_p of -15 volts and -90 volts corresponding to $e\,|\varphi_p|/(kT_{e\delta})$ of 50 and 280, respectively, the experimental data deviated from Eq. (3.50) by factors of about 2 and 6, respectively, as a result of the continuously increasing ion current explained in Chapter I (see Fig. 1-3).

Scharfman and Bredfeldt (1967) proposed that the classical planar space-charge-limited, mobility-controlled diode equations (Cobine, 1958) be employed to locate the sheath edge, especially when $e\,|\varphi_p|/(kT_{e\delta})$ is large such that the sheath is thick. The ion-saturation current then is considered to be the random thermal current at the sheath edge. The concept of this theory is the same as that of Talbot (1960). Talbot, however, was concerned with the collisionless sheath and not the continuum sheath comprising the Scharfman and Bredfeldt's (1967) experiment. Nevertheless, this theory was found to agree with the experimental data of Scharfman and Bredfeldt better than Eq. (3.50) for the high potential limit of $e\,|\varphi_p|/(kT_{e\delta}) = 280$. In contrast, this theory deviated from the experimental data by a factor of about 3 at low probe potentials.

Hoppmann (1966) conducted a shock tube experiment with a parallel double probe in air. Chung (1967) showed that his theory, Eq. (3.50), agrees with Hoppmann's data up to the initial saturation point, with $Sc_i = 1.3$ and $(T_e/T_i)_\delta = 1$.

Thompson (1967) simulated a reentry laminar boundary in an arc-channel for $10^{11} \leq N_{e\delta} \leq 10^{13}$ cm^{-3}. He correlated microwave measurements with electrostatic probe measurements using Chung's equation, Eq. (3.47), except that this equation for the cone was modified so as to apply to a flat plate. The results showed that the probe measurements as interpreted by Chung's theory agreed with the microwave measurements within the experimental scatter of the data.

Several sets of free-flight measurements have been made using continuum probes. In measurements made by Huggins (1972), probes were embedded along the surface of a reentry vehicle near the stagnation region, away from the stagnation region, and in the wake region. For the probes at the stagnation region and along the shoulders of the reentry vehicle, the author essentially employed the vanishing thin-sheath result, Eq. (3.41) with the variables defined by Eq. (3.40) computed according to Newtonian hypersonic aerodynamics. Although no precise examination of the theories can be made from these free-flight tests, Huggins' measurements seemed to show that Eq. (3.41) predicted the electron-number densities satisfactorily. Similar measurements are reported by Hayes and Rotman (1973).

Laboratory experiments with double flush probes were conducted by Seeman and Thornton (1969) in a partially ionized argon flow produced by an electrodeless discharge in a plasma tunnel which yielded charged particle densities in the range of $10^{10} \leq N_{e\delta} \leq 10^{13}$ cm^{-3}. These results also confirm the fact that when the sheath is negligibly thin the ion-saturation current can be computed accurately from the quasi-neutral solution. Seeman and Thornton showed that the formation of foreign deposits on the probe surface distorts the probe *CV* characteristic in a manner such that the electron temperature is grossly overpredicted (see Chapter IV).

In the discussion thus far of probe operations, no distinction had been made between the dimensions of the actual probe surface R_p and that of the body R in which the probe surface is embedded. The distinction is unnecessary when the entire body constitutes the probe, such as the case of a spherical probe in quiescent plasma, or at least when a sufficient surface area surrounding the probe surface is kept at the probe-surface potential. When neither of the above conditions is satisfied, the dimension of the probe surface in relation to that of the body can become one of the governing parameters of the problem. When this situation arises we shall designate the body dimension by R and the probe dimension by R_p. Let us now consider the thick-sheath case.

It should be noted that all the boundary-layer theories discussed thus far assume that convection is negligible within the sheath. Moreover these theories consider that the quasi-neutral boundary layer is at least "locally similar" and that the sheath is one-dimensional. The neglect of convection within the sheath should be quite acceptable so long as the sheath is much thinner than the boundary layer, but convection effects can be expected to become important when the sheath is comparable in magnitude to the boundary-layer thickness. However, if the probe dimension is of the order of or smaller than the sheath thickness, field-fringing effects will become important and the multidimensional aspects of the sheath region must be considered, even if the sheath is thin compared to the boundary layer. We address these questions in the following section.

2. *Thick Sheaths.* As in the quiescent plasma case, thick sheaths occur when $|\chi_p|(\lambda_D/\delta)^2$ is large either because of low electron-number density or large applied potential. For thick sheaths, convective effects in the sheath become important when $ReSc_i \gg 1$.

One of the earliest thick sheath analyses is that of Sonin (1967). He analyzed the ion-collection characteristics of a stagnation probe on a sounding rocket moving at supersonic speeds. The charged particle number density was assumed to be of order $10^3\ \mathrm{cm}^{-3}$ in the ambient atmosphere. The probe radius was taken to be much larger than both the boundary layer and the sheath thicknesses. The electric field was considered sufficiently strong such that $|\chi_p|(\lambda_D/R)^2 \gg 1$ and space-charge effects were neglected. Sonin's one-dimensional treatment of the ion-collection characteristics of his probe yielded ion-current density independent of flow effects, but linearly dependent on the electric field strength.

Space-charge effects must be included in thick sheath analyses for plasmas with charged-particle densities a few orders of magnitude greater than that considered by Sonin. Baum and Denison (1970) solved the compressible boundary layer equations (3.30) and (3.31), simultaneously with the one-dimensional Poisson equation, Eq. (3.33), without introducing any a priori assumptions regarding the division of the boundary layer with a quasi-neutral and a sheath region. Numerical solutions were obtained for sharp conical probes in supersonic flow assuming that the entire surface was at uniform negative potential. Probe currents were found to be, in general, higher than those predicted by thin-sheath theories for the same flow conditions. Later, Baum and Denison (1971) extended the above calculations to include the two-dimensional Poisson equation for the finite-size probe surface comprising the tip of a supersonic cone. The probe surface was assumed to be electrically insulated from the aft portion of the cone, which was kept at a floating potential. The solution of the Poisson equation was obtained in two parts, one for applied potential and one for induced (space-charge) potential. The former was solved as the Laplace equation in two dimensions and the latter as the Poisson equation in one dimension. The two solutions were then linearly superposed. The flow was assumed to be chemically frozen but compressible, and it was assumed that $T_e = T_i$. On the basis of the above model, Baum and Denison showed that the multi-dimensional applied field solution of the Poisson equation dominates if

$$|\chi_p|\left(\frac{\lambda_D}{\delta}\right) \gg \frac{R_p}{\delta} \tag{3.52}$$

and they suggest that the one-dimensional sheath limit is valid if

$$|\chi_p|\left(\frac{\lambda_D}{\delta}\right)^2 \ll 0.005\, R_p/\lambda_D \tag{3.53}$$

Although the results of the theory of Baum and Denison provide information on thick sheath behavior, the theory is too complicated to be used conveniently for probe diagnostics.

Experiments in the thick-sheath regime have been carried out by Dukowicz (1970) and Scharfman and Hammitt (1972). Both experiments were conducted with conical probes of length $L(L = R_p)$ in pressure-driven shock tubes, with the latter experiments designed to vary the parameters $\kappa = Re^{-1}Sc_i^{-1}(L/\lambda_D)^2$ over five decades, i.e., $2 \le \kappa < 10^5$. In Dukowicz's work, the range was $10^{-2} \le \kappa < 10^2$. Figure 3-8 summarizes the above results with a plot of $(j_i L/\lambda_D)(-\chi_p)^{-1/2}$ versus κ. The thick-sheath theories of Baum and Denison (1970) and Hammitt (1970) agree fairly well with each other and lie within the data band as seen in Fig. 3-8; however, they predict a slope with κ considerably less than that indicated by the experiments of Scharfman and Hammitt. The experiments of Dukowicz are in better agreement with these theories over the lower range of κ values. Scharfman and Hammitt suggest the following simple empirical relation for the ion-current density j_i to correlate their data shown in Fig. 3-8.

$$j_i = I/(\pi L^2 \sin\theta)eN_{i\infty}u_\infty = 0.135\,\frac{\lambda_D}{L}(-\chi_p)^{1/2}\kappa^{2/3} \tag{3.54}$$

where θ is the cone half-angle, or

$$j_i(Sc_i Re_L)^{2/3} = 0.135\,\frac{L^{1/3}}{\lambda_D}(-\chi_p)^{1/2} \tag{3.55}$$

However, it may be observed that the scatter in both sets of data shown in

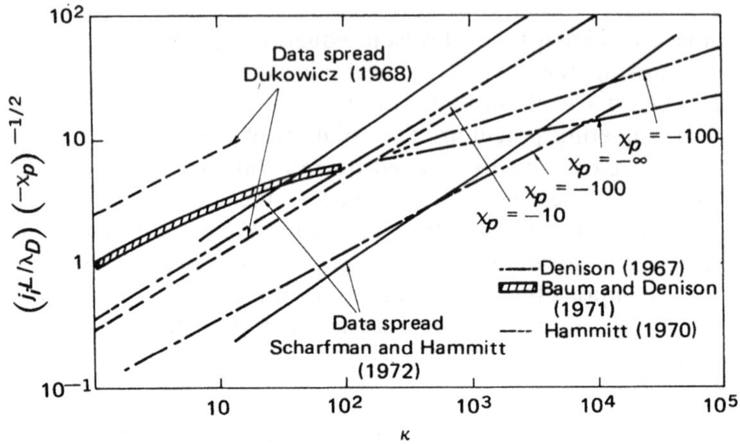

Fig. 3.8. Comparison of experimental results with theories for negatively charged conical probes of length L and vertex half-angle θ in flowing plasmas; $j_i = I/(\pi e u_\infty N_{i\infty} L^2 \sin\theta)$. (From Scharfman and Hammitt, 1972.)

Fig. 3-8 is quite large (because of difficulties associated with shock tube experimentation in the domain of the tests) and for this reason excessive confidence should not be attached to correlations of these data such as Eq. (3.55). At best, what the data seem to show is that thin-sheath theory (i.e., Denison, 1967) better predicts the data trends than do either of the thick-sheath theories plotted in Fig. 3-8. Unpublished experiments by Bruce and Talbot indicate that the thin sheath theories of Denison and Chung and Blankenship [Eqs. (3.47) and (3.49)] do in fact give reasonably accurate predictions of the ion-saturation current to conical probes even for $|\chi_p|(\lambda_D/\delta)^2 > 1$ when the criterion for a thin sheath is no longer satisfied.

The most complete analysis to date for the thick-sheath regime is due to Russo (1972) who obtained numerical solutions to Eqs. (3.1), (3.2), and (3.4) in two dimensions, including finite rate chemistry, for the case of a flush probe on a flat plate. However, in Russo's calculations the ratio T_e/T_i must be specified throughout the flow field. Special attention was paid to the manner in which the three-dimensional outer solution for the electric potential is matched to the two-dimensional inner sheath region. This matching was improperly done by Dukowicz (1970) but handled in a satisfactory approximate manner by Baum and Denison (1971). Because of its general nature, Russo's analysis is capable of handling convective effects in the sheath as well as probe size effects at all probe potentials. Figure 3-9 shows typical three-dimensional contour plots of the charged particle densities obtained from Russo's computations. However, as in the case of the work of Baum and Denison, the Russo analysis requires extensive numerical computation and, therefore, does not conveniently lend itself to application to probe diagnostic work.

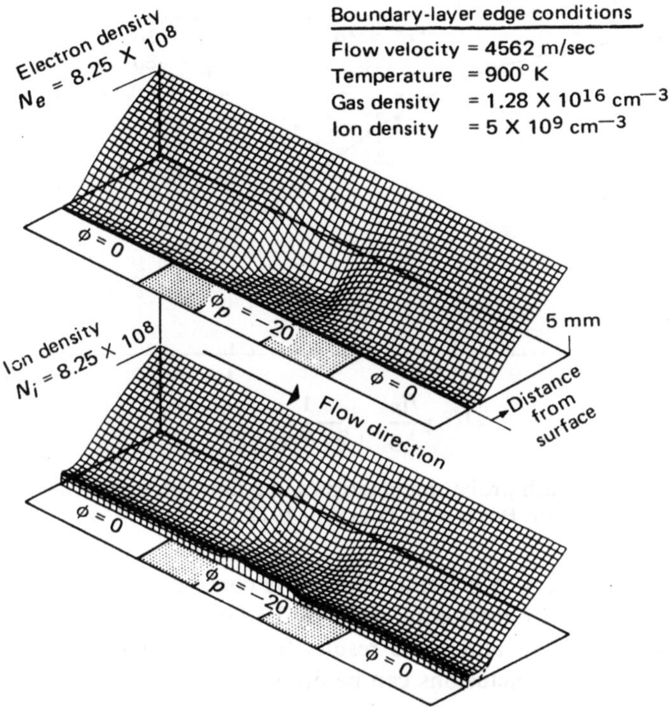

Fig. 3.9. Three-dimensional contour plots of charged particle densities in the neighborhood of a flush probe at −20 V bias. (From Russo, 1972.)

An extensive series of experiments was conducted by Boyer and Touryan (1972) in a hypersonic shock tunnel using flush-mounted electrostatic probes on a sharp flat plate, in which the effects of probe size, geometry, position, bias, and local flow properties were explored in the thick-sheath regime. The range of parameters covered in the tests were $7 < M_\infty < 20$, $10 < Re_{\infty, R} < 10^4$, $1 \leq T_e/T_i \leq 10$, $3 < R/\lambda_D < 4000$, and $-180 < \chi_p < 10$. A correlation was obtained, similar to Eq. (3.55), which relates current density to flow Reynolds number, electron convective flux, probe size R_p and applied potential, as follows:

$$\frac{J_i(Sc_i Re_x)^{1/2}}{N_{e\delta} e u_\delta} \approx 1.8 \left(\frac{\lambda_D}{R_p}\right)^q (-\chi_p)^m \tag{3.56}$$

where

$$0.20 \leq q \leq 0.50; \quad m = [1 + 4\, ln\,(1 + 0.5\sqrt{\kappa})]^{-1} \tag{3.57}$$

and Re_x is the Reynolds number based on the distance from the plate leading edge to the probe location. Figures 3-10 and 3-11 exhibit the data obtained

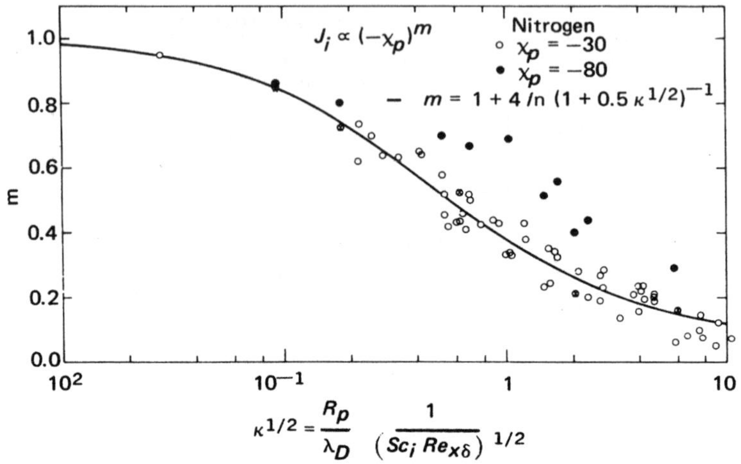

Fig. 3.10. Slope of flush probe characteristics in a nitrogen plasma as a function of the parameter $\varkappa$. (From Boyer and Touryan, 1972.)

by Boyer and Touryan, which yielded the above correlation. Figure 3-11 also shows data from similar tests conducted by Lederman and Avidor (1971); data from Burke (1968), and some results from Scharfman and Bredfeldt (1970). The following conclusions can be drawn.

1. From Fig. 3-11 it can be observed that as R_p/λ_D decreases, the probe current density increases. This effect was also studied systematically by Tseng and Talbot (1970) using flush-mounted probes on a flat plate in continuum flow but under collisionless sheath conditions. It is caused by sheath fringing-field effects on ion-current collection to small probes. Numerical solution of Russo (1972), shown in Fig. 3-12, exhibits the nature of such fringing effects. In Fig. 3-12 equipotential plots for two different-sized probes are compared for the same probe potential.

Figure 3-13 displays the fringing-field data in another fashion as a correlation between current ratio $J_i/J_{i,\mathrm{ref}}$ and the ratio R_p/λ_s, where $J_{i,\mathrm{ref}}$ is the experimentally determined current in the limit $R_p/\lambda_D \gg 1$ and λ_s is the calculated sheath thickness. As can be seen, the fringing-field effect is greater for the case of the collisionless sheath than for the collisional one, because, as one might expect, collisions in the sheath tend to inhibit its three-dimensional growth.

Although the experimental data involve values of $|\chi_p|$ much larger than those upon which the thin-sheath current relationships given by Eqs. (3.41) and (3.50) are based it is interesting to compare the limiting value of J_i of Fig. 3-11 with those given by the thin-sheath theories. Using a reasonable extrapolation, it is seen that the quantity $j\left(\frac{Sc_i}{|\chi_p|}\right)^{1/2}$, where $j = J_{i,\mathrm{sat}}\sqrt{Re_x}/N_{e\delta}eu_\delta$

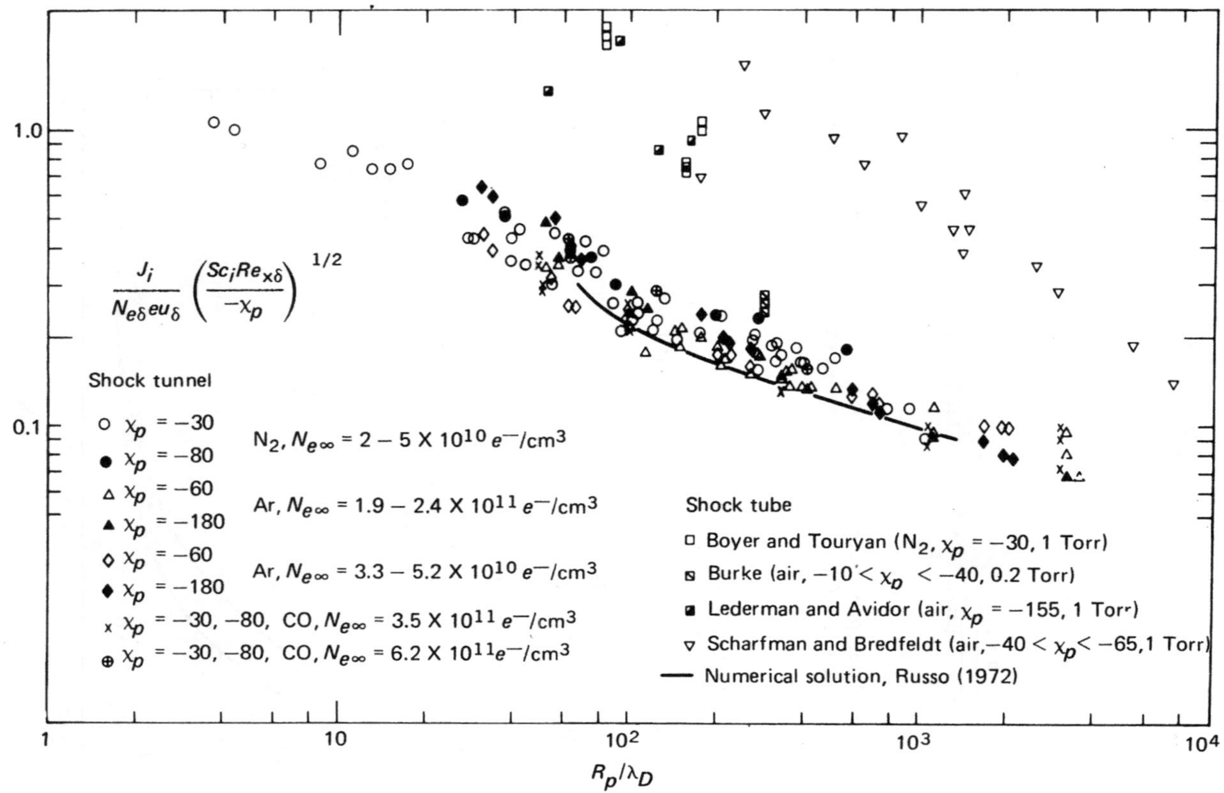

Fig. 3.11. Nondimensional current density to a flush probe in a flowing plasma as a function of the Debye ratio. A composite plot. (From Boyer and Touryan, 1972). Experimental conditions: $7 < M_\infty < 20$; $10 < Re_{(R_p)} < 104$.

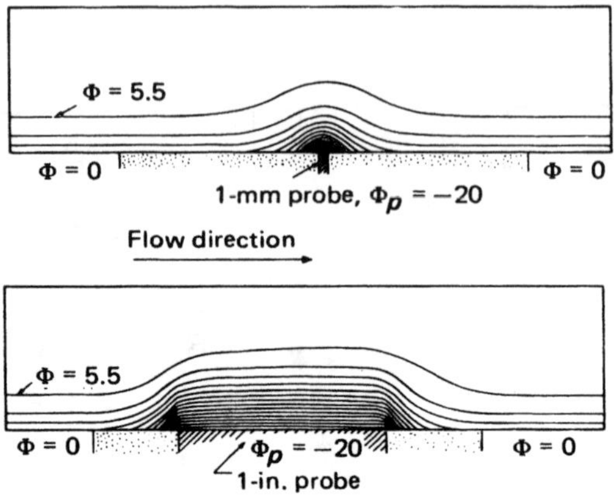

Fig. 3.12. Calculated probe potential contours for flush probes of two different sizes showing fringing effect for conditions same as those in Fig. 3.9. (From Russo, 1972.)

approaches a constant asymptote between 0.05 and 0.08 as $R_p/\lambda_D \to 10^4$. Assuming $Sc_i \simeq 1$, and $|\chi_p| \simeq 80$, j then approaches a value between 0.4 and 0.7. It is interesting to note that the thin sheath solutions [Eqs. (3.41) and (3.50)] give $j = 0.577\sqrt{l}$, and $0.474\sqrt{l}$, respectively, for the thermally equilibrium and frozen flat-plate boundary layers when $(T_e/T_i)_\delta = 1$, and l [cf. Eq. (3.40)] is a quantity of order one.

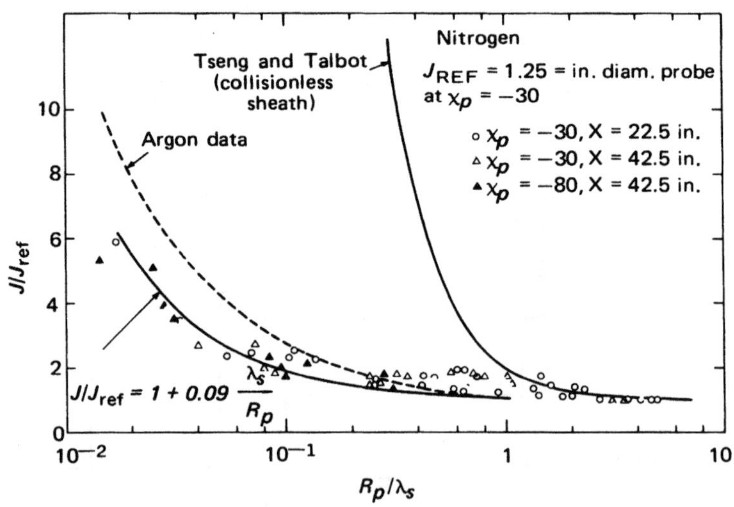

Fig. 3.13. Probe size effect on flush probe ion-current collection in a nitrogen plasma. (From Boyer and Touryan, 1972.)

2. As shown in Fig. 3-10, the current J_i increases with probe potential as $|\chi_p|^m$. In the limit $\kappa \to \infty$, $m \to 0$, which is the thin-sheath limit. At the other extreme, when $\kappa \to 0$, m tends to unity. This is the limit of negligible space-charge effects, and is the case that was studied by Sonin (1967). It can also be shown that the relationships among J_i, χ_p, and $N_{e\delta}$ as given in Eq. (3.56) is consistent with the theoretical results of Stahl and Su (1971).

Experiments in the thick-sheath regime were also conducted by French *et al.* (1970) using spherical probes in a pressure-driven shock tube. The ranges of the parameters covered in these tests were $1.1 \leq M_\infty \leq 1.4$, $4 < Re < 10^4$, $10^{-2} < \lambda_D/R < 0.15$, and $-100 < -\chi_p < -4$, where $Re = \dfrac{2R\rho_\infty u_\infty}{\mu_\infty}$ and R is the probe radius. The data were correlated by the expression

$$\frac{J_i R}{N_{e\infty} e D_{i\infty}} = (-\chi_p)^m (1.2 + 0.09 Re) \tag{3.58}$$

where

$$m = [1 + 0.61\, ln(1 + R/\lambda_D)]^{-1} \tag{3.59}$$

In these experiments the probe sheaths were large compared to the boundary-layer thicknesses, except at the lowest Reynolds numbers, and were comparable with the probe radius. Equation (3.58) was then used to interpret spherical probe data taken in the wakes of hypersonic projectiles. In spite of large differences in Mach numbers between the shock-tube calibration runs and the hypersonic wakes, agreement between probe measurements and microwave measurements of wake electron densities was good. Qualitatively similar results were obtained by Scharfman and Taylor (1971) in an RF plasmajet facility.

C. Probes in Compressible, Finite-Rate-Chemistry Plasma Flows

Few studies are available that deal with continuum electrostatic probes in flowing nonequilibrium plasmas. McAssey and Yeh (1970) carried out a study of compressible stagnation flow. In their study, chemical reactions were limited to the quasi-neutral region. For the highly cooled boundary-layer flow they considered, the finite reaction rate (which was predominantly a recombination reaction with Damkohler number of order unity) was found to result in a reduction in current collection to the probe. Earlier, Chung (1965a) analyzed the effect of nonequilibrium chemistry on a cooled stagnation probe operating in a viscous shock-layer flow, and found a comparable reduction in the current collection rate. Russo's (1972) general solutions yield similar results. For $\mathscr{D} \gg 1$, gas phase recombination reduces the particle gradients at the wall and decreases the probe saturation current, in agreement with McAssey and Yeh and Chung.

3.4 The Collisionless Thin-Sheath (Dense) Case

We come now to the third case mentioned in the Introduction to this chapter. This is the case where $L \gg \lambda \gg \lambda_D$, and this ordering of characteristic lengths describes the situation where the charged-particle motion in the quasi-neutral region of the plasma is governed by continuum phenomena (diffusion, mobility), whereas in the sheath adjacent to a solid boundary the charged-particle motion is collision-free, because the sheath thickness is small compared to the relevant mean free paths.

This case was first analyzed, for a flow situation, by Talbot (1960) in connection with the use of a stagnation-point Langmuir probe for the determination of the free-stream charged-particle density. An improved analysis for this problem was given by Brundin and Talbot (1964). The most recent study of this case is that of Tseng and Talbot (1971), who carried out a combined theoretical and experimental study of the charged-particle density and electron-temperature distributions in a flat-plate laminar boundary-layer flow of a partially ionized gas.

Because this study probably represents the most complete combined experimental and theoretical investigation of the characteristics of surface electrostatic probes in a laminar boundary-layer flow, we shall describe this work in some detail.

In the experiments of Tseng and Talbot point-by-point measurements of n_i and T_e were made throughout the boundary layer, at several streamwise stations, by means of collisionless cylindrical electrostatic probes aligned with the flow. The current-voltage characteristics of flush-mounted surface probes were also obtained at these streamwise stations. One object of the experiment of central importance in the context of the present discussion, was that of establishing the proper interpretation of the current-voltage characteristics of the flush-mounted probes.

In the theoretical investigation, the equation governing $\mathcal{M} = C_i/C_{i\delta}$ for the quasi-neutral portion of the boundary layer was taken to be

$$\left(\frac{\mathcal{M}'}{Sc_A}\right)' + f\mathcal{M}' = 2\frac{d \ln C_\delta}{d \ln s}\left[f'\mathcal{M} + \frac{\dot{w}_i}{\dot{w}_{i\delta}}\frac{\rho_\delta}{\rho}\right] \tag{3.60}$$

where the independent variables are defined in Eq. (3.40) and following Eq. (3.47), with the exponent $\epsilon = 0$. $(\ )'$ denotes the differentiation with respect to η. This equation is derived by combining Eqs. (3.30) and (3.31) with the ion and electron source terms retained, and with the concentration gradient along the boundary-layer edge considered nonzero (but constant). In the derivation the quasi-neutral approximation $C_i/C_{i\delta} = C_e/C_{e\delta}$ is employed, and the T_e/T_i gradient contribution to the electron diffusion is neglected since it is small compared to the concentration gradient term. The $\rho\mu$ ratio l was taken as constant through the boundary layer, in order to make use of existing solutions for the stream function f and the neutral gas-density distribution ρ/ρ_δ and temperature distribution $T_n/T_{n\delta}$. However, the ambipolar Schmidt number $Sc_A = (\mu/\rho)\, D_A$ [where $D_A = D_i(1 + \hat{\tau})$] was treated as a variable, and the variable quantity $\hat{\tau} = T_e/T_i$ was determined from boundary-layer solutions for $T_i = T$, and measurements of T_e. Using experimentally determined values of the boundary-layer edge conditions, and employing the Hinnov–Hirschberg (1962) expression for the volume recombination terms, Eq. (3.60) was integrated subject to the boundary condition $\mathcal{M}(0) = \mathcal{M}_s$ and $\mathcal{M}(\infty) = 1$, $\mathcal{M}'(\infty) = 0$. $\mathcal{M}_s$ was chosen to be consistent with the flush probe measurements of the ion density at the sheath edge, explained in detail below.

For the calculation of the electron-temperature variation through the boundary layer, the electron-energy equation Eq. (3.32) is employed with the T_e/T_i gradient contribution to the diffusion flux neglected but $\dot{w}_e h_e$ and $\dot{w}_{he}$ terms of Eq. (3.3) retained. The boundary conditions for the electron-energy equation are specified at the outer edge of the free-fall sheath and at the boundary-layer edge. Following the treatment of Dix (1964), a thin transitional layer is postulated to exist between the sheath and the quasi-neutral region. The net electron energy flux crossing the outer edge of the transitional layer is equated to the net energy flux crossing the edge of the free-fall sheath. As shown by Dix, this can be expressed by

$$\left.\frac{\partial\theta}{\partial\eta}\right|_s = g_s J_{i,\text{wall}}\, \theta_s^{-3/2} \tag{3.61}$$

where $\theta = T_e/T_{e\delta}$, and $g_s = 2(\chi_f - \frac{1}{2})l^{1/2}(\rho_\delta/\rho_s)(x/Re_x^{\frac{1}{2}})(k/e\ K_{he})_s$ with χ_f being the dimensionless potential drop across the sheath.

The results of the calculations are compared with the experimental data in Fig. 3-14, for the station 4 in. downstream from the plate leading edge. It can be seen that agreement is excellent for the ion density distribution through the boundary layer, and quite good for the electron-temperature distribution.

The ion flux received by a flush probe, extrapolated back to its value at the floating potential, can be used to determine the ion-number density at the sheath edge through application of a collisionless sheath theory of Bienkowski (1967). Bienkowski's analysis yields the expression

$$(N_i)_{\text{sheath edge}} = \hat{\tau}_s \frac{J_{i,\text{wall}}(m_i/2kT_e)_s^{1/2}}{(1 + \hat{\tau}_s)A_o(\hat{\tau}_s)} \tag{3.62}$$

where $J_{i,\text{wall}}$ is the measured ion-number density flux at the floating potential, and $A_o(\hat{\tau})$ is a function of order unity obtained by equating the two expressions for number density given by Eqs. (4.9) and (4.11) in Bienkowski's paper, assuming the wall emission terms to be zero and neglecting a term of order $(m_e/m_i)^{1/2}$ [and correcting an obvious misprint in Eq. (4.9)]. Equation (3.62) is, in fact, a modified form of the Bohm sheath criterion.

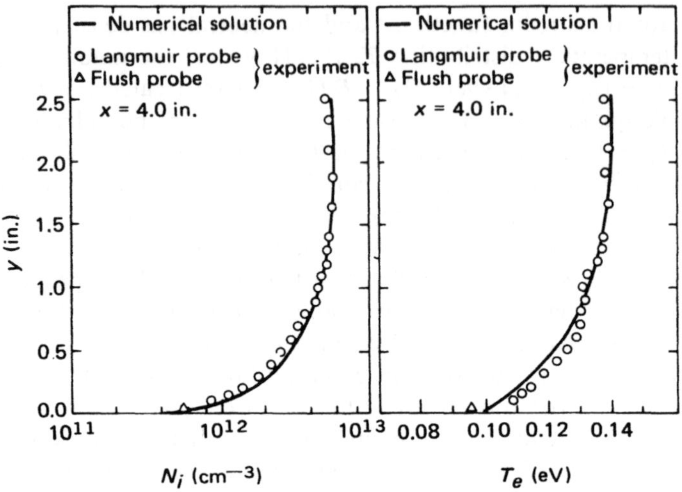

Fig. 3-14. Comparison between theoretical and experimental charged-particle density and electron-temperature profiles in an argon plasma boundary-layer flow. (From Tseng and Talbot, 1971.)

The number density at the sheath edge obtained in this fashion is represented in Figure 3-14 by the triangular data point plotted at approximately $y = 0$. It will be noted that this point falls very nicely on the theoretical curve and the extrapolation of the data obtained within the boundary layer.

It is of interest to compare the ambipolar diffusion flux values obtained from the flush-probe measurements with those obtained by theory and those

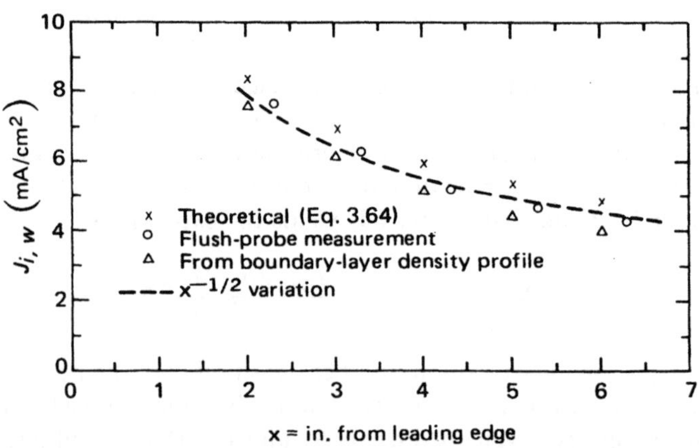

Fig. 3.15. Comparison between theory and experiment for ambipolar and diffusion flux to flush probes in an argon plasma boundary-layer flow. (From Tseng and Talbot, 1971.)

found by using the concentration gradient evaluated at one mean free path from the wall, as determined by the slope of the experimental ion density profile within the boundary layer, according to

$$J_{i,w} = \frac{e\rho D_A}{m_i} \frac{\partial C_i}{\partial y}\bigg|_{y=\lambda} \tag{3.63}$$

These three values of $J_{i,w}$, given in units of current density $\mathrm{m}A/\mathrm{cm}^2$, are shown as a function of distance from the leading edge in Fig. 3-15. It can be seen that agreement among the three values is very good.

This good agreement between theory and experiment establishes the fact that flush-probe measurements can be used to determine the charged particle density at the boundary-layer edge, at least under the conditions comprehended by the theory and experiment under consideration. In fact the theoretical values of $J_{i,w}$ given in Fig. 3-15 are related, according to the theory, to the boundary-layer edge ion concentration $(N_i)_\delta$ by

$$J_{i,w} = (N_i)_\delta \frac{eRe_x^{1/2}}{x} \left\{ D_A \left(\frac{\rho}{\rho_\delta}\right)^2 \frac{\mathcal{M}'(\eta)}{2l^{1/2}} \right\}_{\eta=\eta_\lambda} \tag{3.64}$$

where η_λ is the value of η at $y = \lambda$. In general, $\mathcal{M}'(\eta_\lambda)$ is not known a priori, for it must be calculated for a chosen value of $\mathcal{M}_s$, and $\mathcal{M}_s$ is not known until $N_{i\delta}$ is known. This difficulty can be overcome by starting with an estimate for $\mathcal{M}_s$, and thus obtaining a first approximation to $N_{i\delta}$. Then using Eq. (3.62) for N_{is}, an improved value of $\mathcal{M}_s$ can be found. Because $\mathcal{M}'(\eta_\lambda)$ depends only weakly on $\mathcal{M}_s$, a convergent iterative process can be established easily.

3.5 Summary

The governing parameters and basic behavior of steady-state continuum probes are well established for probes consisting of perfect conductors when there are no surface emissions or fluid turbulence.

The present status of the theory of continuum probes is such that the electron density and temperature at the edge of the diffusion layer can be determined from the ion-saturation current and from the current-voltage sweep between the floating potential and the ion-saturation potentials, provided that certain conditions discussed in this chapter are satisfied. Most efforts have been directed toward analyzing the characteristics of the spherical probe in a quiescent plasma and the flush-mounted probe in a plasma boundary layer. The status of these analyses with respect to their application to practical plasma diagnostics is summarized below.

1. Spherical Probe in Quiescent Plasma

Detailed analyses of the probe characteristics are available for constant-property plasmas. However, in practice, a considerable cooling of the heavy gas and a corresponding variation of the plasma properties take place through the diffusion layer. No general solutions for a variable-property plasma are available that can be employed for diagnostic purposes. For $T_e = T_i$, however, the constant-property solutions for the ion-saturation current can be employed to determine the electron concentration in the undisturbed plasma, $N_{e,\infty}$, if the chemical reactions involving the ionized species are frozen and if the sheath is thin. This is because the governing equations for the variable-property case can be transformed into those for the constant-property case by assuming that $\rho\mu$ is constant.

This transformation results in the solution for the dimensionless ion-saturation current, $j_{i,\text{sat}}$, for the variable-property case, which is the same as that for the constant-property case except that the ion-diffusion coefficient D_i appearing in the definition of $j_{i,\text{sat}}$ (see Fig. 3-4) is changed to $D_{i,\infty}l$, where l is given by Eq. (3.40). Of course, the value of χ_p at which the saturation is attained varies, but this is unimportant. According to the Baum and Chapkis (1970) solution shown in Fig. 3-4, the initial ion saturation is attained at $\chi_p \simeq -10$ for $(R/\lambda_{D,\infty})^2 \gtrsim 10^3$, and $j_{i,\text{sat}}$ is a little greater than 2. The initial saturation point and the value of the saturation current are more clearly recognizeable if the CV relationship is plotted on a linear or semilog scale as explained in Chapter I.

As will be noted subsequently, $j_{i,\text{sat}}$ for a boundary layer with frozen T_e is within about 20 percent of that with equilibrium T_e when $T_{e,\delta} = T_{i,\delta}$, the actual difference being a weak function of Sc_i. It is therefore logical to assume that such is also the case in a quiescent plasma, hence that the measurement of $j_{i,\text{sat}}$ for a spherical probe in a constant property plasma with $T_e = T_i$ can be employed with reasonable accuracy to determine $N_{e,\infty}$ for both equilibrium and frozen electron temperatures provided that $T_{e,\infty} = T_{i,\infty}$. However, when $T_{e,\infty} \neq T_{i,\infty}$, the use of the existing constant-property solutions, which among other things assume that T_e/T_i is constant, is questionable. T_e/T_i is not constant through the diffusion layer, and the use of the simple logarithmic relationship to determine $T_{e,\infty}$ cannot be substantiated.

For the thick sheath case nonnegligible with $\chi_p(\lambda_D/R)^2$, the current to a spherical probe in a variable-property plasma is expected to vary in the same general manner as shown by the constant-property solutions such as those shown in Fig. 3-4. However, the detailed CV relationship, will depend on the property variations, and use of thick sheath solutions for diagnostics is not a simple matter.

2. Flowing Plasmas: Probes Large Relative to Sheath Thickness

General and correlated solutions are available which can be employed to determine the $N_{e,\delta}$ and $T_{e,\delta}$ when the chemical reactions involving the ionized species are frozen in the boundary layer, and the conditions of Eq. (3.39) are satisfied.

When $T_e = T_i$, Eq. (3.41) can be used to determine $N_{e,\delta}$ from the measured saturation current for both two-dimensional and axisymmetric boundary layers.

When the electron temperature is frozen with respect to the ion temperature in the boundary layer, Eqs. (3.50) and (3.51) may be used to determine $T_{e,\delta}$ and $N_{e,\delta}$. The current-voltage sweep as well as the ion-saturation current are needed to determine these values when $T_{e,\delta} \neq T_{i,\delta}$. Equation (3.50) is again applicable to the types of the boundary layers for which Eq. (3.41) is applicable. Equation (3.50) is constructed by correlating the various numerical solutions, and therefore, is not as accurate as Eq. (3.41).

When $T_{e,\delta} = T_{i,\delta}$, Eq. (3.50) for the frozen T_e gives values of $j_{i,\text{sat}}$ which are within about 20 percent of those for equilibrium T_e given by Eq. (3.41) when $Sc_i \approx 1$. In contrast, for $(T_e/T_i)_\delta \gg 1$, values of $j_{i,\text{sat}}$ given by Eq. (3.50) increase linearly with $(T_e/T_i)_\delta$.

When the peak electron concentration is within the diffusion boundary layer, and when it can be assumed that the chemical reaction is frozen between the location of peak electron concentration and the probe, an equation such as (3.47) can be constructed from Eq. (3.41) to determine the peak concentration when $T_e = T_i$. Equation (3.47) is for the boundary layer over a cone. However, such equations can be readily constructed for any boundary layer for which Eq. (3.41) is applicable. Relationships such as Eq. (3.47) can be employed to determine the peak electron concentration in certain chemically reacting boundary layers.

3. Flowing Plasmas: Probes Small Relative to Sheath Thickness

A few specific solutions have been obtained, but no general solutions are available that can be employed to determine $N_{e,\delta}$ and $T_{e,\delta}$. The correlation of the various flat-plate experimental data shown on Figs. 3-10 and 3-11, guided by the solutions of Russo and Touryan (1972), yield Eq. (3.56). Until further analyses become available, one may, for instance, write Eq. (3.56) in the following explicit form, using mean values for the exponents q and m

$$\frac{J_{i,\text{sat}}}{eN_{e,\delta}u_\delta}\sqrt{Re_{x,\delta}} = (1.8)\left(\frac{\lambda_{D,\delta}}{R_p}\right)^{0.4}\left(\frac{-\chi_p}{Sc_i}\right)^{0.5} \tag{3.65}$$

Additionally, if use is made of the known general behavior of boundary layers, one may put Eq. (3.65) into the following form applicable to the more general class of boundary layers for which Eq. (3.41) is valid.

$$\frac{J_{i,\text{sat}}}{eN_{e,\delta}N_{e,\delta}}\frac{\sqrt{s/l}}{r^\epsilon\mu_\delta} = (1.8)\left(\frac{\lambda_{D,\delta}}{R}\right)^{0.4}\left(\frac{-\chi_p}{Sc_i}\right)^{0.5} \tag{3.66}$$

Equation (3.66) is only an extrapolation of Eq. (3.65), and its validity is not substantiated.

There is as yet no satisfactory theory for boundary-layer probes operating in the thick sheath regime, $|\chi_p|(\lambda_D/\delta)^2 \gg 1$. The available experimental evidence suggests that thin-sheath theories such as those represented by Eqs. (3.41) or (3.50) in fact provide more accurate predictions for the relationship between probe current and free-stream electron density than do the several thick-sheath theories that have been proposed.

4. Classical Space-Charge Sheath Equation

The classical high-pressure sheath equation, the diode equation, along with the ion thermal-current expression is at times employed for probe data reduction [see Scharfman and Bredfeldt (1970), Hayes and Rotman (1973)]. The basic shortcomings of such an approach are as follows.

a. It neglects the partial-pressure diffusion of the charged particles.
b. It assumes that the ion current to the probe is given by the ion thermal current at the sheath edge.
c. Usually, no self-consistent method is used to match the sheath solution to the quasi-neutral region.

These approximations seem unnecessarily crude in view of the present understanding of continuum plasma probe characteristics.

For sufficiently thin sheaths, the ion-saturation current is largely, if not entirely, governed by the quasi-neutral diffusion layer. Therefore whatever sheath theory one uses, sufficiently accurate values of $N_{e,\delta}$ should be obtainable provided that the quasi-neutral region is analyzed correctly.

Scharfman and Bredfeldt (1967) employed both Eq. (3.50) and the simple theory based on the diode equation. For the case of $\chi_p = O(1)$, wherein the sheath was thought to be thin, the diode theory gave the values of $N_{e,\delta}$, which differed from their experimental values by factors of 2 to 3, whereas Eq. (3.50) agreed very closely with the experimental results. This disagreement of the diode theory is evidently due to the fact that the approximations associated with the diode theory are unsatisfactory for describing the quasi-neutral region. For the larger values of χ_p employed, presumably producing thick sheaths, the diode theory gave the variations of j_i with respect to χ_p, which were three to four times more rapid than those predicted by the continuum solution of Stahl and Su (1971) given in Fig. 3-6. Although the Stahl and Su solution is for a constant-property plasma, it is rather doubtful that compressibility effects alone are responsible for the large difference.

5. Use of the Electron Saturation Current for Diagnostics

According to the theory for continuum probes in a constant-property plasma, there exists an electron-saturation current similar to the ion-saturation current. Therefore, in principle it should be possible to use the electron-saturation current, as well as the ion-saturation current, to determine $N_{e,\delta}$, and this in fact is true for probes operating in the collisionless regime. In practice, however, in continuum probes the electron-saturation current is not as clearly observed as the ion-saturation current, and its relation to the electron density is rather complicated for a variable property plasma. Chung (1965a) has argued that as the probe surface becomes biased to draw more electron than ion current, a positive potential is generated outside of the diffusion layer to satisfy this net current requirement. This positive potential heats the incoming electrons in the inviscid region, thus increasing the electron pressure and mobility, which in turn results in increased electron diffusion to the probe surface. The greater electron current to the probe, in turn, requires a greater positive potential in the inviscid region. Therefore use of a positive probe potential can result in an amplified perturbation of the electron temperature of the inviscid region. This perturbation can extend a considerable distance into the plasma because of the extremely high electron thermal conductivity and anomalies associated with the relationship between the electron current and the plasma properties ensue.

References

Barad, M. S. and Cohen, I. M., Univ. of Pennsylvania Rep. 2281/10, (1973) Philadelphia, Pennsylvania. See also *Phys. Fluids*, **17**, April 1974.

Baum, E. and Chapkis, R. L. (1970), *AIAA J.*, **8,** 1073.

Baum, E. and Denison, M. R. (1970), TRW Rep. 06488-64556-R0-00, September 1970.

Baum, E. and Denison, M. R. (1971), TRW Rep. 06488-6526-R0-00, September 1971.

Bienkowski, G. K. (1967), *Phys. Fluids*, **10,** 381.

Boyd, R. L. F. (1951), *Proc. Phys. Soc.* (*London*), **B64,** 795.

Boyer, D. W. and Touryan, K. J. (1972), *AIAA J.*, **12,** 1667.

Brundin, C. L. and Talbot, L. (1964), AGARD Rep. 478. AGARD meeting on Arc Heaters and MHD Accelerators for Aerodynamic Use. Rhode-St. Géniese, Belgium.

Burke, A. F. (1968), AIAA Paper No. 68-166, 6th Aerospace Sciences Meeting, New York, 1968.

Burke, A. F. and Lam, S. H. (1967), AIAA Paper No. 67-100. 5th Aerospace Sciences Meeting, New York, 1967. See also Burke, A. F., Cornell Aeronautical Laboratory Rep. No. AN-2101-Y-1 May 1967.

Bush, W. B. and Fendell F. E. (1970), *J. Plasma Phys.*, **4,** 317.

Carrier, G. F. and Fendell, F. E. (1970), *Phys. Fluids*, **13**, 2966.

Chapkis, L. R. and Baum, E. (1971), *AIAA J.*, **9**, 1963.

Chung, P. M. (1962), TDR-169 (3230-12) TN-2, Aerospace Corporation, San Bernardino, California.

Chung, P. M. (1964), *Phys. Fluids*, **7**, 110.

Chung, P. M. (1965a), *AIAA J.*, **3**, 8817.

Chung, P. M. (1965b), *Advances in Heat Transfer*, Volume 2, Chap. 2, Academic Press, New York.

Chung, P. M. (1967), *J. Spacecraft Rockets*, **4**, 1105.

Chung, P. M. (1969), *Phys. Fluids*, **12**, 1623.

Chung, P. M. and Blankenship, V. D. (1966a), *AIAA J.*, **4**, 442.

Chung, P. M. and Blankenship, V. D. (1966b), *J. Spacecraft Rockets*, **3**, 1715.

Chung, P. M. and Mullen, J. F. (1963), AIAA Paper No. 63-161, 1st Aerospace Sciences Meeting, New York, 1963.

Cicerone, R. and Bowhill, S. (1967), Univ. of Illinois, Rep. AR-21, Urbana, Illinois.

Cobine, J. D. (1958), *Gaseous Conductors*, Dover, New York.

Cohen, I. M. (1963), *Phys. Fluids*, **6**, 1492.

Cohen, I. M. (1967), *AIAA J.*, **5**, 63.

Cohen, I. M. (1970), *Phys. Fluids*, **13**, 889.

Cohen, I. M. and Schweitzer, S. (1968), *AIAA J.*, **6**, 298.

David, T. S. (1971), Ph.D. Thesis, 1971, Cornell University, Ithaca, New York.

deBoer, P. C. T. and Johnson, R. A. (1968), *Phys. Fluids*, **11**, 909.

Denison, M. R. (1967), TRW Rep. 06488-06065-R0-00, Redondo Beach, California.

Dix, D. M. (1964), Aerospace Corporation Rep. ATN-64-(9232)-1, El Segundo, California.

Dukowicz, J. K. (1969), CAL Rep. No. RA-2641-Y-1, Buffalo, New York.

Dukowicz, J. K. (1970), CAL Rep. No. AN-2755-Y-1, Buffalo, New York.

French, I. P., Hayami, R. A., Arnold, T. E., Steinberg, M., Appleton, J. P., and Sonin, A. A. (1970), *AIAA J.*, **8**, 2207.

Hammitt, A. G. (1970), TRW Rep. 06488-6433-R0-00, Redondo Beach, California.

Hayes, D. T. and Rotman, W. (1973), *AIAA J.*, **11**, 675.

Hinnov, E. and Hirshberg, J. C. (1962), *Phys. Rev.*, **125**, 795.

Hoppmann, R. F. (1966), Renselaer Polytechnic Inst. Rep. TRAE 6605, Albany, New York.

Hoult, D. P. (1965), *J. Geophys. Res.*, **70**, 3183.

Huggins, R. W. (1972), AIAA Paper No. 72-691, AIAA Fluid and Plasma Dynamics Conference, Boston, Massachusetts, June 1972.

Inutake, M. and Kuriki, K. (1972), 8th Rarefied Gas Dynamics Symposium, Stanford University, California.

Johnson, R. A. and deBoer, P. C. T. (1972), *AIAA J.*, **10**, 664.

Jou, W. H. and Cheng, S. I. (1971), *Phys. Fluids*, **14**, 2144.

Kiel, R. E. (1969), *J. Appl. Phys.* **40**, 3668.

Kulgein, N. G. (1968), *AIAA J.*, **6**, 151.

Lam, S. H. (1964), *AIAA J.*, **2**, 256.

Lam, S. H. (1965), AIAA Paper No. 65-543, San Francisco, September 1965.

Lam, S. H. (1968), Proc. 8th Int. Conf. Phenomena of Ionized Gases, Vienna, Austria, August 1967, p. 545.

Lederman, S. and Avidor, J. (1971), *Israel J. Tech.*, **9,** 19.
Little, R. G. and Waymouth, J. F. (1966), *Phys. Fluids*, **9,** 801.
McAssey, E. V., Jr. and Yeh, H. (1970), *J. Heat Transfer*, **92,** 447.
Persson, K. B. (1962), *Phys. Fluids*, **5,** 1625.
Russo, A. J. (1972), Sandia Laboratories Rep. No. SC-RR-72-0111, Albuquerque, New Mexico.
Russo, A. J. and Touryan, K. J. (1972), *AIAA J.*, **12,** 1675.
Scharfman, W. E. and Bredfeldt, H. R. (1967), Standard Research Institute Final Rep. Project 6138, Menlo Park, California.
Scharfman W. E. and Bredfeldt, H. R. (1970), *AIAA J.*, **8,** 662.
Scharfman, W. E. and Hammitt, A. G. (1972), *AIAA J.*, **10,** 434.
Scharfman, W. E. and Taylor, W. C. (1971), Standard Research Institute Rep. Project No. 7712, Menlo Park, California.
Seemann, G. R. and Thornton, J. A. (1969), AIAA Paper No. 69-700, 2nd Fluid and Plasma Dynamics Conference, June 1969.
Sonin, A. A. (1967), *J. Geophys. Res.*, **72,** 4547.
Stahl, N. and Su, C. H. (1971), *Phys. Fluids*, **14,** 1366.
Su, C. H. (1965), *AIAA J.*, **3,** 842.
Su, C. H. and Kiel, R. E. (1966), *J. Appl. Phys.*, **37,** 4907.
Su, C. H. and Lam, S. H. (1963), *Phys. Fluids*, **6,** 1479.
Talbot, L. (1960), *Phys. Fluids*, **3,** 289.
Tanaka, T. and Hirao, K. (1972), Report of Ionospheric and Space Research in Japan, **26,** p 121.
Thomas, D. L. (1969), *Phys. Fluids*, **12,** 356.
Thompson, W. P. (1967), Aerospace Corporation Rep. No. TR-0158(340-20)-4, San Bernardino, California.
Toba, K. and Sayano, S. (1967), *J. Plasma Phys.*, **1,** 407.
Touryan, K. J. and Boyer, D. W. (1972), Proc. IUTAM Symposium on Dynamics of Ionized Gases, Tokyo, Japan, September 1971.
Tseng, R. C. and Talbot, L. (1971), *AIAA J.*, **9,** 1365.
Whitman, A. M. and Chien, C. J. (1971), *Phys. Fluids*, **14,** 1115.
Zakharova, V. M., Kagan, Y. M., Mustafin, K. S., and Penel, V. I. (1960), *Zh. Tekhn, Fiz.*, **30,** 442.

CHAPTER IV

Special Topics

4.0 Introduction

This chapter will be devoted to several topics on electrostatic probes for which our understanding is not as complete as that of the subjects discussed previously. An attempt will be made to review the present state of knowledge of these topics.

We will discuss five special topics of interest in the following order—(a) probe-surface phenomena, (b) laminar plasmas with negative ions, (c) strongly ionized plasmas, (d) turbulent plasmas, and (e) plasmas in magnetic fields.

4.1 Probe-Surface Phenomena

In our discussion of electric probes thus far we have dealt almost exclusively with cold conductors that have fully catalytic surfaces. At such surfaces, all positive ions immediately recombine with electrons and the neutralized species then return to the plasma leaving a negligible concentration of ionized species at the surface. If the surface potential is more positive than the floating potential, the electron-current flux to the surface from the plasma exceeds the ion-current flux. The excess electrons are then immediately absorbed by the surface, since a conductor, by definition, is a perfect absorber of electrons. This net electron current flux into the conductor then constitutes the positive electric current. In contrast, when the surface potential is below the floating potential the positive ion flux from the plasma exceeds that of the electrons. The excess ions are neutralized by the electrons conducted through the solid to the surface. The net electrons supplied to the surface by the conductors constitute the negative electric current.

On a probe surface that is not fully catalytic, the recombination process is not immediate as compared to the rates at which the ionized species arrive at the surface. Therefore depending on the relative rates of recombination, ionized species transport, etc., nonnegligible concentrations of ions and electrons exist at the surface. These concentrations are not known *a priori* but result as a solution of the plasma equations satisfying the nonequilibrium surface kinetic relationship.

When a probe surface is heated to a sufficiently high temperature in relation to the work function, thermionic emission of electrons will occur. The thermionic process is usually much more efficient than the transport processes of the ionized species in the plasma. Hence the nonnegligible surface concentration of electrons may be considered as that given by the equilibrium thermionic emission relations. The surface concentration of ions is governed by the surface-catalytic properties and the availability of electrons as discussed earlier.

At higher temperatures and large potentials, surface emission of ions can also take place. Moreover, the high-energy ionized species arriving at the surface can cause secondary emissions from the surface. There are many other surface phenomena that can occur depending on the plasma and surface conditions and materials, but these will not be discussed here.

Often, condensible gases such as ablation products in the plasma result in deposition of a surface layer. Surface deposits can readily alter the catalytic properties and electrical conductivity of the probe surface as well as the other properties mentioned above. As we can see from the above discussion, this can completely alter the probe characteristics. Failure to recognize the presence of surface layers in the reduction of probe data could result in completely erroneous plasma diagnostics. In fact surface-layer deposits constitute one of the most vexing and serious problems in probe operation.

However, because of the inherent complexities associated with surface-layer phenomena and the wide variety of types of surface layers on probes in different environments, a general treatment of the problem is not possible. Indeed, there are no systematic studies either theoretical or experimental that deal with all the surface phenomena discussed above. Very few studies consider probe surfaces other than fully catalytic and nonemitting. Experimental data are equally scarce. A few of the more important effects of the surface phenomena on the probe characteristics will be elaborated below.

4.1.1 Effect of Surface Deposit Layers

Let us consider surface-layer deposits first. Seemann and Thornton (1969) have obtained some interesting experimental data on contaminated probes. Their study was aimed at identifying the magnitude of contamination problems on airborne (e.g., reentry vehicle) planar double probes. Figure 4-1 illustrates the serious effect of ablation products on flush-mounted double probes. Note how markedly the current-voltage characteristics are distorted when the probe surface becomes covered with insulating deposits. Figure 4-1 also shows the change in the shape of the *CV* characteristics with increasingly shallower slopes accompanying the increased surface contamination. Figure

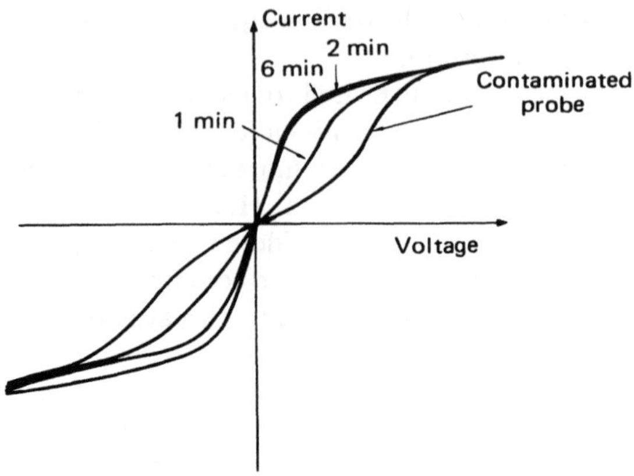

Fig. 4-1. *CV* characteristic of contaminated probe. The time referred to on curves is the time that probe was operated as a cathode at a current of 1 mA for the purpose of cleaning the probe surface prior to obtaining the *CV* characteristic. (From Seemann and Thornton, 1969.)

4-2 shows similar results from Starner's (1966) experiment in which he used a flat-faced Langmuir probe in an argon arc facility. Note the sharp decrease in the *CV* slope near the floating potential—a change that would lead to an erroneously large estimate of T_e [see also Hirao and Oyama (1972)].

Deposits of insulating materials on *portions* of the electrode surface will reduce the effective electrode surface area. Conversely, deposits of conducting

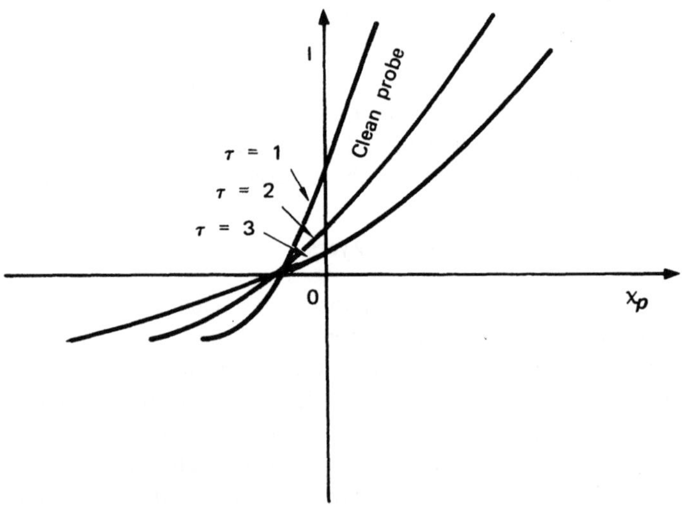

Fig. 4-2. Characteristic of contaminated probe as a function of time probe remains in plasma. τ-Relative time scale. (From Starner, 1966.)

material, originating from evaporation or sputtering of the electrode surface and deposited on the insulator surfaces adjacent to the electrodes, can increase the effective area of the electrodes as well as cause leakage currents.

A partially catalytic or a non-catalytic surface would produce the same effect as a film deposit, normally a decrease in the saturation current to the probe. This decrease is effected through a decrease in charged particle gradients at the wall (or at the sheath edge). Detailed discussions of absorbing, emitting, and reflecting surfaces are given by Burke (1967) and Sajben (1970).

Changes in the work function of the probe surface are less detrimental to the probe characteristics. A study by Medicus (1961) for probes with non-uniform work functions indicates that this situation resulted in a rounding of the knee of the plane probe characteristic but does not seriously affect the slope of that part of the characteristic from which T_e is estimated.

A method that has been found useful in minimizing the effects of contamination is to maintain the probe at a large negative potential, of the order of 100–200 V, and to superimpose on it a voltage pulse that carries the probe voltage through the range of the characteristic. The large negative voltage produces sufficiently energetic ion bombardment of the probe surface to sputter off contaminants, thus providing a clean surface prior to recording the probe characteristic. This technique has been used by many investigators, e.g., Waymouth (1959) and Peterson (1968), and is discussed also by Swift and Schwar (1971).

Another factor of concern in probe experiments is that of plasma depletion or drain. Generally speaking, it is preferable to operate probes in the ion-attracting mode rather than the electron-attracting domain because the current drawn and hence the perturbation of the plasma is much less in the case of ion collection. In a plasma at rest, negligible charged particle drain by the probe may be translated into the condition that the current drawn by the probe be much smaller than the total rate of depletion of charged particles by all other loss mechanisms. In the case of a flowing plasma, the criterion for negligible plasma drain is roughly that the convective flux of charged particles across a representative cross section of the flow must be much greater than the current extracted from the flow by the probe. What constitutes a representative cross section will vary with the conditions of the experiment, and each experimental situation must be evaluated on its own merits. The same is true of problems involving the perturbation of the plasma properties by probe supports, lead wires, insulators, and the like. Some of these considerations are discussed by Swift and Schwar (1971).

4.1.2 Effect of Electron Emission

Emission of electrons can greatly perturb the potential distribution in the neighborhood of an electrostatic probe and thus invalidate theories that do

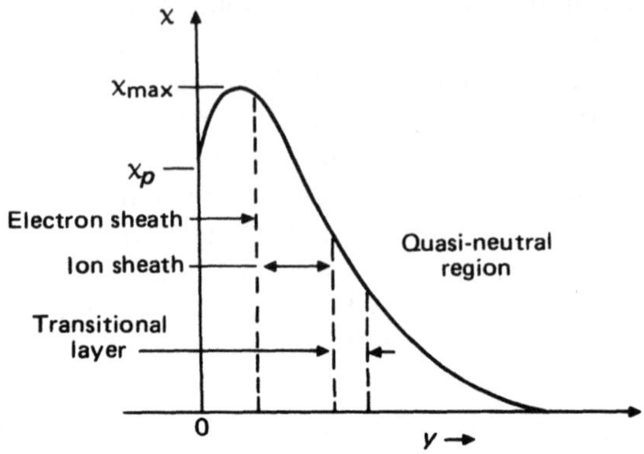

Fig. 4-3. Schematic potential distribution for an emitting probe with thick double sheath. (From Chang and Bienkowski, 1970.)

not take emission currents into account. The electron emission can be thermionic emission in the case of hot probes or secondary emission resulting from ion bombardment. Several studies are available that deal with electron emission from probe surfaces in the collisionless limit. Bienkowski and Kalnavarns (1968) carried out an investigation of the effects of surface boundary conditions on sheath structure and *CV* characteristics in the continuum limit. They considered the emitted particle fluxes as freely specifiable boundary conditions. The most striking result of their analysis is the prediction of a sharp increase in apparent ion-saturation current to the probe caused by emitted electrons escaping into the plasma.

Chang and Bienkowski (1970) conducted a systematic investigation of the effects on probe response caused by electrons emitted from the probe surface with a half-Maxwellian distribution. The analysis dealt with a spherical probe in the collisionless, transitional and the dense regimes. The results obtained were not unlike those found for the continuum case, but the electron emission was found to be space-charge limited for the collisionless and transitional regimes. For the dense plasma with a collisionless sheath the electron emission was found to be both space-charge and diffusion limited. Chang and Bienkowski give detailed studies of the sheath structure for space-charge limited emission. Figure 4-3 is a schematic potential distribution for the thick, double sheath that results from space-charge limited operation of the probe. The space-charge limitation is evidenced by the appearance of a potential maximum χ_{max} intervening between the probe and the undisturbed plasma.

The Chang and Bienkowski results show that for thermionic emission, the space-charge limitation phenomenon at small and moderate $|\chi_p|$ invalidates all conventional methods of extracting information about the electron temperature and space potential. Although the electron-saturation current is generally unaffected, the observed or apparent ion-saturation current is increased by the emission. When the level of emission is relatively low so that the apparent ion current is not restricted by the space-charge limitation condition, the electron temperature can be obtained in the usual manner if the Chang and Bienkowski results are used to separate the electron current from the apparent ion current. This, for example, is the case with secondary emission currents caused by ion bombardment on metallic probes.

In practice, the level of electron emission is not known and there exists no simple method of determining electron temperature or the space potential from the measured *CV* characteristics. However, the Chang and Bienkowski results can be used as a guide for intelligent interpretation.

It is interesting to note that the increase in apparent ion-saturation current caused by thermionic emission could account for anomalous *CV* characteristics observed under some flight conditions where the two saturation current branches of the *CV* curves are found to be nearly equal in magnitude $\left(\text{instead of in the ratio } \sqrt{\frac{m_i T_e}{m_e T_i}}\right)$. Another possible cause of symmetric characteristics is the presence of negative ions, to be discussed in the next section.

4.2 Negative Ions

Touryan and Chung (1971) formulated the governing equations for a continuum constant property plasma including negative ions. The resulting equations were then solved for a flat-plate boundary-layer flow, under the assumption that the layer could be represented by a thin sheath and a quasi-neutral region. Subsequently Bailey and Touryan (1973) numerically integrated the coupled species conservation and Poisson equations without making the *a priori* division of the boundary layer into two regions.

Figure 4-4 shows typical profiles of the charged particle number densities $\tilde{N}_+$ and $\tilde{N}_-$ for the quasi-neutral region normalized with respect to $\tilde{N}_{+0}$ where subscript 0 denotes the boundary-layer edge. For $\tilde{N}_{-0} = 0.9$ case the sheath profile is also shown. Figure 4-5 exhibits the sheath structure for several values of negative ion concentration and electron-to-heavy particle temperature ratio τ. Note the slight decrease in sheath thickness as the edge negative ion concentration $\tilde{N}_{-0}$ is increased.

As is shown in Fig. 4-5, the quasi-neutral diffusion layer becomes thicker and the positive ion slope near the wall increases as the concentration of the negative ions is increased. Indeed, Fig. 4-6 shows that the presence of negative ions suppresses the electron-saturation current and increases slightly the positive-ion current. This increase is more pronounced as τ increases (increasing T_e or decreasing T_i). The effect of decreasing the negative-ion mobility further increases the positive saturation current and decreases the negative

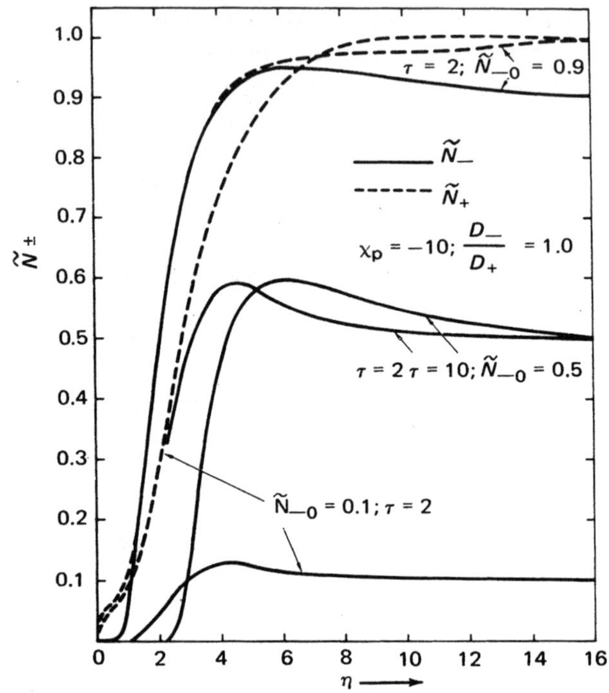

Fig. 4-4. Charged-particle density profiles over flush probe in the presence of negative ions including the effect of changing electron–ion temperature ratio τ. (From Touryan and Chung, 1971.)

saturation current. These results are in good qualitative agreement with the experimental measurements of Starner (1969) who used a flowing argon plasma with various quench gases, such as SF_6 and SF_4, added to the plasma stream. Inflight electrostatic probe measurements by Hayes *et al.* (1974) with freon as the electrophilic additive show similar qualitative agreement with the above predictions.

Also, we may note from Fig. 4-4 that the $\tilde{N}_-$ profiles show rather unexpected peaks within the boundary layer. These peaks are the result of convective effects that cause the negative ions to "pile up" against a strong repulsive field. In the absence of convection, the peaks in the negative-ion concentrations disappear and the corresponding reduction in electron concentration is not observed. The peaks also decrease when the applied field becomes positive and the probe is operated in the negative-charge collection mode.

The governing equations for a continuum plasma with negative ions consist of the conservation equations for the ions, the electrons and the negative ions, together with the Poisson equation. The conservation equations for

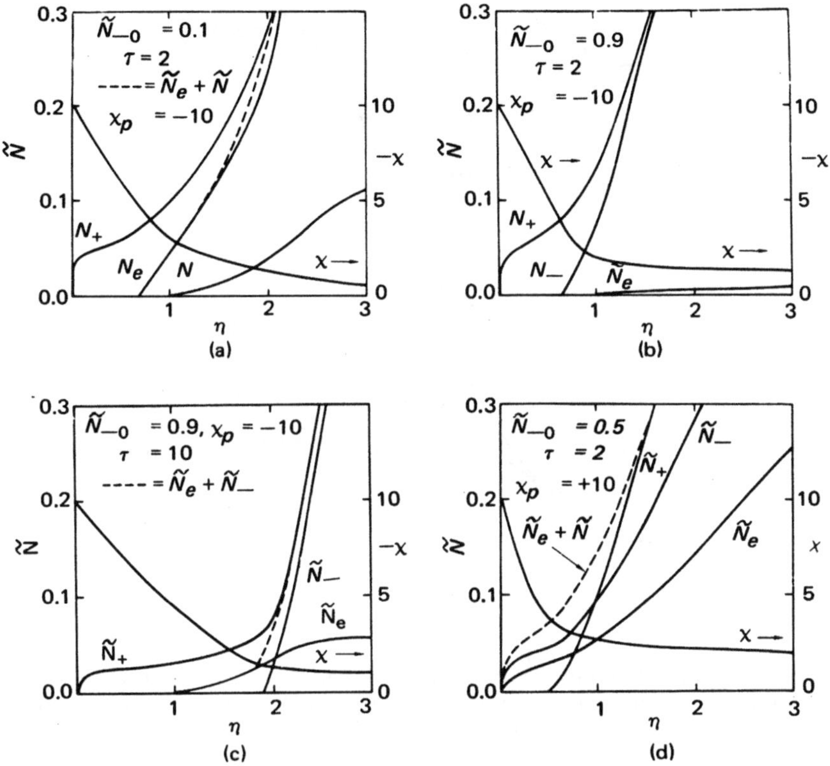

Fig. 4-5. Sheath structure over flush probe in the presence of negative ions. (From Bailey and Touryan, 1973.)

the positive ions and the electrons are the same as those discussed in Chapter III. The conservation equation for the negative ions is the same as that for electrons except for the different values of diffusivity and mobility. The Poisson equation, however, contains an additional term, and takes the form

$$\nabla \cdot \vec{E} = \frac{e}{\sigma}(N_+ - N_- - N_e) \tag{4.1}$$

Hence in the quasi-neutral region we have

$$N_+ - N_- - N_e = 0 \tag{4.2}$$

The three conservation equations can be combined with the above condition of quasi-neutrality, to yield for the incompressible quasi-neutral region,

$$\vec{V} \cdot \nabla \left(\frac{N_+}{D_+} + \frac{N_-}{D_-} + \frac{N_e}{D_e} \right) = \nabla^2(N_+ + N_- + N_e) \tag{4.3}$$

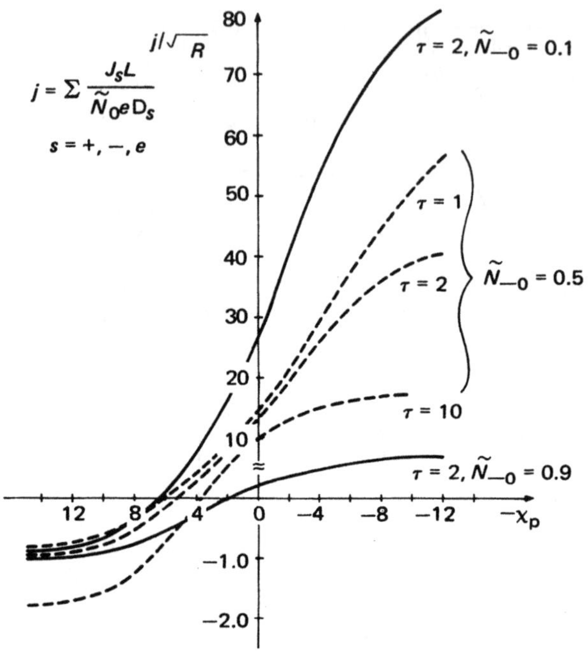

Fig. 4-6. Current-voltage characteristics for various values of negative-ion concentrations and electron–temperature ratios. (From Touryan and Chung, 1971.)

where $\vec{V}$ is the velocity vector. Now, when $N_- = 0$ and $N_e = N_+$ this equation reduces to the usual quasi-neutral equation

$$\vec{V} \cdot \nabla N_+ = \frac{2}{\left(\dfrac{1}{D_i} + \dfrac{1}{D_e}\right)} \nabla^2 N_+ \tag{4.4}$$

When $N_- \neq 0$ and $\vec{V} \neq 0$, Eq. (4.3) cannot be reduced to an equation such as Eq. (4.4) with one dependent variable. Instead even in the quasi-neutral region Eq. (4.3) must be analyzed together with the other conservation equations and Eq. (4.2). Thus when $\vec{V} \neq 0$, the negative ions couple directly with the other charged species, and fundamentally alter the electrical structure of the boundary layer, even in the quasi-neutral region.

In contrast the negative ions are not coupled with the other charged particles when $\vec{V} = 0$. This is seen readily from Eq. (4.3). When $\vec{V} = 0$, Eq. (4.3) reduces to

$$\nabla^2 N_+ = 0 \tag{4.5}$$

$$\nabla^2 N_- = 0 \tag{4.6}$$

$$\nabla^2 N_e = 0 \tag{4.7}$$

The conservation equations for the positive ions and the electrons in the case for which there are no negative ions present also reduce to Eqs. (4.5) and (4.6), respectively, when $\vec{V} = 0$. Therefore the coupling between the negative ions and the other charged particles is through the convection, and this was shown by Touryan and Chung (1971) to be the cause for, among other things, the peaks in N_- observed within the quasi-neutral boundary layer.

Luzzi and Jenkins (1971) studied the characteristics of collisionless Langmuir probes in the presence of negative ions. They considered the particle current to the probe in the attracting field to be equal to its thermal current whereas that in the repulsive field to be equal to the thermal current multiplied by exp $(-\chi)$. Because convection was not considered, there was no coupling between the negative ions and the other charged particles. Luzzi and Jenkins derived a relationship that enables one to determine the depletion of electrons in a plasma caused by the introduction of an electron-attaching gas, from knowledge of the positive saturation current and the difference between the floating potentials before and after the introduction of the electron-attaching gas.

4.3 Strongly Ionized Plasmas

As is discussed in the Appendix, a weakly ionized plasma is defined as a plasma wherein the collisions between the ionized species are negligible as compared to those between the ionized and neutral species. Because the Coulomb cross sections for collisions between the ionized species are about two orders of magnitude larger than the ionized–neutral and neutral–neutral cross sections, the characterization of a plasma as weakly ionized usually implies a plasma with the degree of ionization much less than one percent, although to be more precise one should examine the criterion given by Eq. (A.50).

Plasmas that do not satisfy the above criterion are usually called moderately or strongly ionized plasmas. It is obvious from the collision criterion that the overall momentum and energy equations are coupled to the momentum and energy equations of the ionized species in a moderately or strongly ionized plasma. This causes the analysis of moderately and strongly ionized plasmas in the continuum flow regime to be much more complicated than that of a weakly ionized plasma.

Su and Sonin (1967) treated the problem of a spherical electric probe in a collision-dominated, moderately ionized, isothermal plasma. Although they relaxed the assumption that charged particles collide with neutrals only, they used diffusion coefficients that depended only on neutral particle density, and were thus constant. Their results show that, in the limit $R/\lambda_D \to \infty$, the solution of the moderately ionized case for moderately negative potentials can be obtained from that for the weakly ionized case [e.g., Cohen (1973)] by a simple transformation.

Barad and Cohen (1973) have developed a theory and obtained numerical solutions for a spherical probe in a collision-dominated, quiescent plasma in which the ratio charge–charge to charge–neutral collision frequencies ν_{ei}/ν_{en} may range from zero (weakly ionized plasma) to unity (strongly ionized plasma). The Su and Sonin (1967) restriction of constant T_e is eliminated by introducing the electron-energy equation and retaining the diffusion term in the electron-current equation. However, following Jou and Cheng (1971) Barad and Cohen assume the ion temperature to be constant and equal to T_n, the neutral gas temperature. They neglect the collisional coupling terms in the electron-energy equation [see Eq. (A.75)] between electrons and ions (or neutrals), which cannot be wholly justified for a dense plasma with a collisional sheath. Nevertheless, Barad and Cohen obtain some interesting results that show that charge–charge collisions have a significant effect on the electron-attracting branch of the CV characteristics, as shown in Fig. 4-7. In fact, as the ratio ν_{ei}/ν_{en} is raised from zero to unity, the calculated nondimensional electron saturation current increases sharply. For the ion-

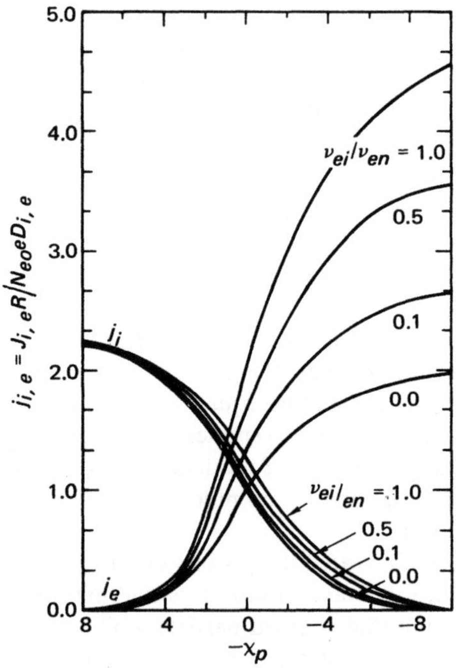

Fig. 4-7. Dimensionless electron and ion currents versus nondimensional probe voltage, for various values of the ratio ν_{ei}/ν_{en}. (From Barad and Cohen, 1973.)

collection branch of the characteristics, the corresponding increase is much less pronounced.

Numerical sheath calculations by Barad and Cohen show that the values of probe potential ranging from strongly positive to slightly negative ($\chi_p = 10$ to -2), the effect of charge–charge interactions is to create a decrease in the normalized electron-number density in the plasma surrounding the probe. The net result is that the plasma near the probe has a more positive potential than it would have had in the weakly ionized limit.

The above analysis is of limited usefulness, however, in that when the degree of ionization in a plasma is high, λ_D is almost always much smaller than the characteristic mean free path, the sheath is collisionless, and therefore the sheath theories discussed in Chapters II and III (Section 3.4) are adequate. It is then the analysis of the quasi-neutral region that distinguishes the analysis of the strongly ionized plasma from that of the weakly ionized plasma in a continuum flow. In contrast, the analysis of the collisionless flow regime is the same as that for the weakly ionized plasma in the same regime.

Johnson and Touryan (1967) and Shelton *et al.* (1968), and Shelton (1970) carried out a detailed derivation of the governing equations for the strongly

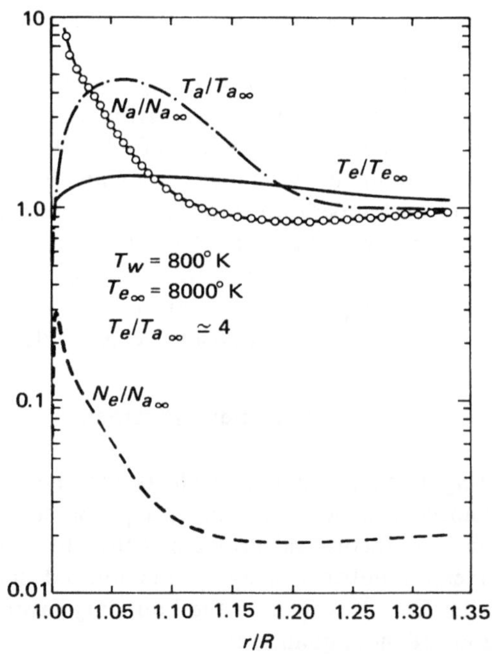

Fig. 4-8. Electron and heavy particle flow profiles for the stagnation shock layer flow of a strongly ionized plasma, $M_\infty = 2.5$, $Re_\infty = 85$. (From Shelton *et al.*, 1968.)

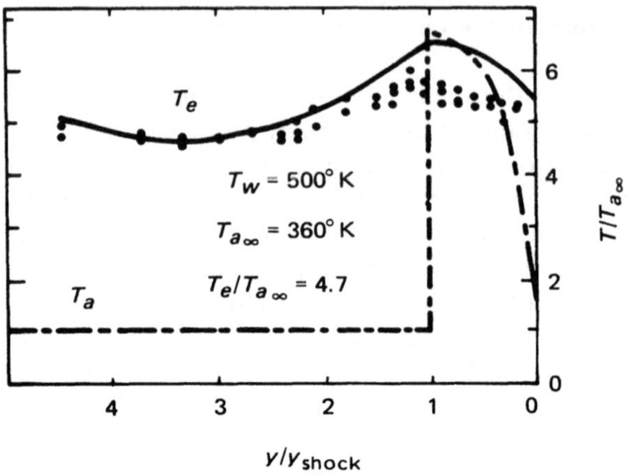

Fig. 4-9. Comparison between measured and predicted electron temperatures for the stagnation flow of a strongly ionized plasma. (From Sugimoto and Nishida, 1972.)

ionized variable property plasmas in continuum flow. These equations were solved for the stagnation region of an axisymmetric blunt body. As expected, these governing equations reduce to those of the weakly ionized plasmas as the degree of ionization decreases.

Figure 4-8 shows the structure of a strongly ionized viscous shock layer in thermal nonequilibrium obtained by solving the multifluid equations, mentioned above, for the stagnation region of a blunt probe. It is interesting to note the large decoupling between the electron and heavy-particle temperatures, with the electron temperature showing some precursor effect.

Similar calculations were made by Nishida (1972) and verified experimentally by Sugimoto and Nishida (1972) for somewhat lower ionization levels. The thermal coupling between electrons and heavy particles is somewhat stronger for the Sugimoto and Nishida case, as depicted in Fig. 4-9.

4.4 Turbulent Plasmas

The fluctuating instantaneous as well as the mean values of electric current and potential can be measured at a probe surface immersed in a turbulent plasma. The theoretical problem is then to relate these measured values to the mean number density of the ionized species and electron temperature in the free stream or in the boundary layer and to the mean-square fluctuations of these quantities.

For probes operating in the collisionless regime, the instantaneous *mechanisms* of collecting or repelling the ionized species are independent of the turbulent properties of the neutral gas. Hence the only basic difference between turbulent and laminar probe theories is that in the former one must now consider the unsteady boundary conditions imposed at the free-stream as compared to the steady boundary conditions in the laminar case. However, this does not mean that the problem is simple to analyze. The complexity arises from the fact that the unsteady conditions imposed by the free stream are random conditions that must be treated by stochastic analysis. Of course, these random conditions of the ionized species are closely tied to those of the neutral gas.

For probes operating in the continuum regime, however, the instantaneous mechanisms of the electron and ion-particle and energy transports are directly coupled to the turbulence properties of the neutral gas in the viscous layer. Therefore the difference between the continuum turbulent and laminar probe theories is more fundamental than that for the collisionless probes.

Because of the complexities of the problem, published works on the subject of electrostatic probe diagnostics of turbulent plasmas are few in number. Also, the theory is at a quite preliminary stage of development. This is more so for the case of continuum probes than for collisionless probes. Therefore we shall first discuss the case of collisionless probes for which the theory is more complete.

4.4.1 Collisionless Regime

As has already been mentioned, the instantaneous mechanisms of collection and repulsion of the ionized species by a probe in the collisionless regime is independent of the neutral gas turbulence because there are no collisions within the region of influence of the probe.

The forerunner to the problem of probe theory in a turbulent plasma is that of probe response in a plasma containing harmonic fluctuations. In this regard, Barnes and Eros (1950), Garscadden and Emeleus (1962), Sugawara and Hatta (1963), and Crawford (1963), have attempted to predict the average probe current for known harmonic fluctuations in the electron density and temperature and the space potential of the plasma. Berman and Lam (1968) made a detailed analysis of the relationship between the various quantities measured at the probe surface and the harmonic oscillations of the ionized-species concentrations in the ambient plasma. Among other things the latter study showed that the response of the probe to the fluctuations of the plasma is practically instantaneous provided that the fluctuation frequency of the plasma is below a critical frequency, which is somewhat less than the ion plasma frequency $\omega_{pi} = (e^2 N_e/m_i\sigma)^{1/2}$. Therefore the unsteady plasma–probe relationship can be constructed as a succession of instantaneous steady-state relationships; that is, a quasi-steady approximation can be made if the fluctuation frequency is sufficiently less than the plasma frequency.

In most engineering flow systems, the important frequencies associated with hydrodynamic turbulence are much lower than the plasma frequency. Taking advantage of this and of the fact that the instantaneous electron- and ion-transport mechanisms are unaffected by turbulence, Demetriades and Doughman (1966), Peterson (1968), Sutton (1969), and Peterson and Talbot (1970) have analyzed probe response in a turbulent plasma stream through the use of available steady-state solutions.

Demetriades and Doughman (1966) considered the orbital-motion-limited case (see Chapter II), that is the case for which λ_D/R is large compared to unity. The electron Mach number, that is, the ratio of the flow velocity to the speed of sound in the "electron gas," is very small for most plasmas because of the extremely small electron mass. Therefore the electron current to the probe is independent of the free-stream plasma velocity. The starting point of the analysis of Demetriades and Doughman (1966) is the following set of the steady-state electron-current relationships derived by Langmuir and Mott-Smith (1926) for quiescent plasmas, which have been referred to earlier in Chapter II.

$$J_e(N_e, T_e, \varphi) = \text{const}\, N_e T_e^{1/2} e^{e\varphi/(kT_e)} \qquad \text{for} \quad \varphi \leqq 0 \tag{4.8}$$

$$J_e(N_e, T_e, \varphi) = \text{const}\, N_e T_e^{1/2} \left(1 + \frac{e\varphi}{kT_e}\right)^{\alpha} \qquad \text{for} \quad \varphi \geqq 0 \tag{4.9}$$

In the above equations, J_e, N_e, and T_e are the electron-current density, number density, and temperature, respectively; φ is the probe potential with respect to the plasma; and the exponent α depends on the probe geometry. It is 0, $\frac{1}{2}$, or 1 for the plane, cylindrical, or spherical probe. Equation (4.9) is valid only when the normalized probe potential $|e\varphi/kT_e|$ is large compared to unity (see Chapter II).

Each variable in Eqs. (4.8) and (4.9) is first divided into a mean ($^-$) and a fluctuating Δ component,

$$N_e(t) = \bar{N}_e + \Delta N_e(t),$$
$$T_e(t) = \bar{T}_e + \Delta T_e(t) \tag{4.10}$$

etc.

Then when $J_e(N_e, T_e, \varphi)$ is expanded in a Taylor series, about the computed value $J_e(\bar{N}_e, \bar{T}_e, \bar{\varphi}) = J_{e,0}$ and averaged, the following expression for the measured mean current $\bar{J}_e$ results.

$$\frac{\bar{J}}{J_0} = 1 + \left(\frac{J}{J_0}\right) [f_{NT}\, \overline{\Delta N \Delta T} + f_{N\varphi}\, \overline{\Delta N \Delta \varphi} + f_{NT}\, \overline{\Delta N \Delta T}]$$
$$+ \frac{1}{2J_0} [f_{NN}\, \overline{(\Delta N)^2} + f_{TT}\, \overline{(\Delta T)^2} + f_{\varphi\varphi}\, \overline{(\Delta \varphi)^2}] \tag{4.11}$$

where the subscript e has been discarded for convenience, and terms of order

Δ^3 and higher have been neglected under the assumption that the turbulent fluctuations are sufficiently small. The symbols f_{NT}, $f_{N\varphi}$, etc., denote partial derivatives of $J_e(N_e, T_e, \varphi)$ given by Eqs. (4.8) and (4.9) evaluated at $\bar{N}_e$, $\bar{T}_e$, and $\bar{\varphi}$.

Similarly, the general expression for the measured mean-square fluctuations of the current density is obtained as

$$\overline{(\Delta J)^2} = f_N f_N \overline{(\Delta N)^2} + f_\varphi f_\varphi \overline{(\Delta\varphi)^2} + f_T f_T \overline{(\Delta T)^2} + 2f_N f_T \overline{\Delta N \Delta T} + 2f_N f_\varphi \overline{\Delta N \Delta\varphi} + 2f_T f_\varphi \overline{\Delta T \Delta\varphi} \tag{4.12}$$

By measuring the values of $\bar{J}$ and $\overline{(\Delta J)^2}$ for several values of the applied mean potential $\bar{\varphi}$, and by evaluating the partial derivatives f_N, f_{TT}, etc., at these mean potentials, a set of equations can be constructed. These equations, with the help of certain other auxiliary relationships, can be solved by iteration for some of the correlation functions such as $\overline{(\Delta N)^2}$ thus relating the measured values to the turbulent plasma properties.

Demetriades and Doughman (1966) showed through such analyses that a cylindrical probe, aligned in the flow direction, can be most conveniently employed in the highly positive electron-saturated ion-current range to yield the plasma quantities J_0 and $\overline{(\Delta N)^2}/\bar{N}^2$ from the measured quantities $\bar{J}$ and $\overline{(\Delta J)^2}$ according to

$$\operatorname{Lim} \frac{\bar{J}}{J_0} = 1 \quad \text{as} \quad \frac{e\bar{\varphi}}{k\bar{T}} \to \infty \tag{4.13}$$

and

$$\operatorname{Lim} \frac{\overline{(\Delta J)^2}}{J_0^2} = \frac{\overline{(\Delta N)^2}}{\bar{N}^2} \quad \text{as} \quad \frac{e\bar{\varphi}}{k\bar{T}} \to \infty \tag{4.14}$$

Furthermore, Demetriades and Doughman (1966) showed that the actual mean electron temperature, $\bar{T}$, of the turbulent plasma can be computed, to an accuracy of order $\overline{(\Delta T)^2}$, directly from Eq. (4.8) according to

$$\bar{T} = \frac{e}{k} \frac{d\bar{\varphi}}{d \ln \bar{J}} \tag{4.15}$$

Note that this is precisely the equation that gives the electron temperature of a laminar plasma except that φ and J are replaced by their averaged values.

Demetriades and Doughman verified at least qualitatively their analysis by an experiment conducted in a turbulent argon plasma jet.

Sutton (1969) studied in detail the applicability of cylindrical probes to the measurement of the turbulence characteristics of hypersonic turbulent plasma wakes generated in ballistic ranges. In particular, Sutton examined the ballistic range conditions under which the collisionless probe theories could be used, and studied the effects of collisions on the interpretation of the data. In addition, various mechanical and electrical circuit aspects of probe operation are discussed in the Sutton paper.

Peterson (1968) and Peterson and Talbot (1970) carried out an analysis similar to that of Demetriades and Doughman (1966), but their analysis used cylindrical single and double probes operating in the regime of $\lambda_D/R \lesssim 0.2$ in contrast with that of Demetriades and Doughman, which was for $\lambda_D/R \gg 1$. For their regime, Peterson and Talbot (1970) employed the steady-state solutions of Laframboise (1966) as approximated by Eq. (2.17).

A comparison of Eq. (2.17) with Eq. (4.9) shows that J_e for $\lambda_D/R \lesssim 0.2$ involves T_i as well as T_e, whereas J_e for $\lambda_D/R \gg 1$ is independent of T_i. Additionally, the dependence of $J_{e,i}$ on N_e and T_e is nonlinear, which complicates the analysis.

The Taylor series expansion of $J_e(N_e, T_e, T_i, \varphi)$ about $J_{e0}(\bar{N}_e, \bar{T}_e, \bar{T}_i, \bar{\varphi})$ up to terms of second order, and subsequent averaging yields an equation similar to Eq. (4.11) except that it now contains additional terms involving $f_{NT_i}, f_{T_eT_i}, f_{T_i\varphi}$, etc. Similarly, an equation for $\overline{(\Delta J)^2}$ was derived by Peterson and Talbot (1970), which is analogous to Eq. (4.12) except that it also contains additional terms. Corresponding equations for the double probe were also derived. These equations for the double probes do not contain the correlation terms associated with fluctuations in the plasma potential, and hence, are substantially simpler than the corresponding single-probe equations. However, by constructing various partial derivatives form Eq. (2.17), Peterson (1968) showed that for the moderately large values of the non-dimensional potential, $\chi_p = e\varphi/(kT_e)$, the partial derivatives $f_{N_e\varphi}, f_\varphi$, etc., involving the potential, approach zero for the single probes. For large values of χ_p, then, the equations for the single probes describing J_e and $\overline{(\Delta J_e)^2}$ become identical in form to those for the double probes. Furthermore when the ion–electron temperature ratio of the free-stream plasma, $\bar{T}_i/\bar{T}_e$, is very small, the partial derivatives involving the ion temperature become also negligible. Therefore in the limit of moderately large χ_p and small $\bar{T}_i/\bar{T}_e$, Peterson and Talbot (1970) reduced their equations to the following forms.

$$\frac{J}{J_0} = 1 + \frac{1}{2J_0}\left[f_{NN}\frac{\overline{(\Delta N)^2}}{\bar{N}^2} + f_{TT}\frac{\overline{(\Delta T)^2}}{\bar{T}^2} + f_{TT}\frac{\overline{\Delta N \Delta T}}{\bar{N}\bar{T}}\right] \tag{4.16}$$

$$\frac{\overline{(\Delta J)^2}}{J_0^2} = \left(f_N f_N \frac{\bar{N}^2}{J_0^2}\right)\frac{\overline{(\Delta N)^2}}{\bar{N}^2} + \left(f_T f_T \frac{\bar{T}^2}{J_0^2}\right)\frac{\overline{(\Delta T)^2}}{\bar{T}^2} + 2\left(f_N f_T \frac{\bar{T}}{J_0^2}\right)\frac{\overline{\Delta N \Delta T}}{\bar{N}\bar{T}} \tag{4.17}$$

where the subscript e has been discarded for convenience.

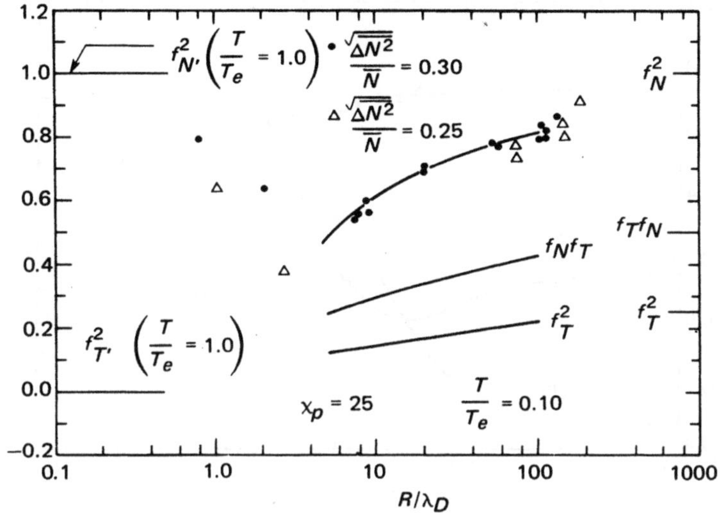

Fig. 4-10. Normalized mean-square saturation electron current to a cylindrical probe function of R/λ_D. [See Eq. (4.17)]. (From Peterson and Talbot, 1970.)

The above equations apply to both single and double probes. The partial derivatives f_N, f_{NT}, etc., are evaluated from Eq. (2.17) for single probes, and from the analogous double-probe equations for double probes.

Peterson and Talbot (1970) conducted an experiment in a turbulent argon plasma jet using single and double cylindrical probes. By analyzing the experimental data in light of their theory, they showed among other things that for $|\chi_p| \gtrsim 20$, and $T_i/T_e = 0.1$, the last two terms of Eq. (4.17) are negligible as compared to the first term on the right-hand side of that equation (see Fig. 4-10 for single-probe data and Fig. 4-11 for double-probe data). Therefore the mean-square fluctuation of the current measured at the probe is a direct measure of the mean-square fluctuation of the electron density in the free stream. Recall that this was also the case with the probe operating in the regime of $\lambda_D/R \gg 1$ [see Eq. (4.14)].

Further reductions of Eq. (4.16) as well as other relationships pertaining to the electron temperature and floating potential fluctuations were also given in Peterson (1968).

Before leaving the topic of collisionless probe response in turbulent plasmas, it may be pointed out that both Demetriades and Doughman and Peterson and Talbot employed cylindrical probes aligned in the flow direction, which are incapable of responding to the eddies whose sizes are smaller than, say, one-half of the probe length. In contrast, it can be inferred from Koopman's (1971) studies of transient flow that a cylindrical probe oriented

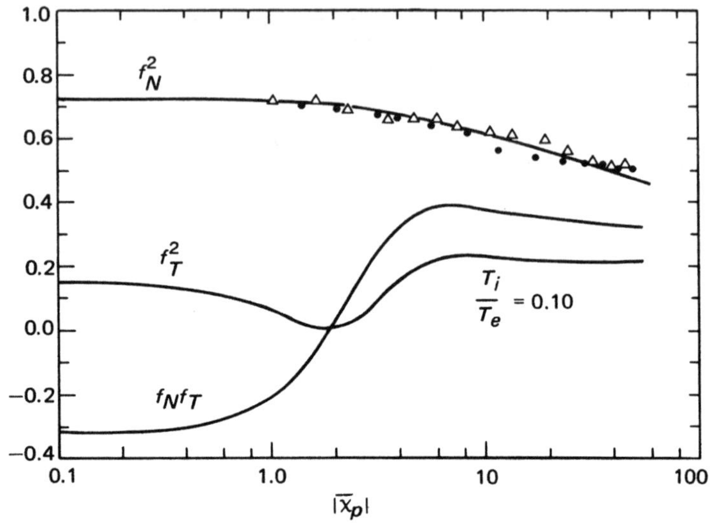

Fig. 4-11. Normalized mean-square current collection by a double cylindrical probe with $R/\lambda_D = 7.5$ and $\frac{\sqrt{\overline{(\Delta J)^2}}}{J^2} = 0.27$ [see Eq. (4.17)]. Here $|\bar{\chi}_p|$ denotes the normalized potential difference between the two probes. (From Peterson and Talbot 1970.)

transverse to the flow can respond to the eddies whose size are of the order of the probe diameter. It can be further inferred from Koopman's work that the theory necessary to interpret the turbulent plasma data for an ion-attracting cylindrical probe oriented transverse to the flow may be more direct and simpler than the theories for electron- or ion-attracting cylindrical probes aligned with the flow as discussed here.

4.4.2 Continuum Regime

In a high Reynolds number flow, the electric communication between the free-stream plasma and the continuum probe takes place across a hydrodynamic boundary layer. Because in many cases of practical interest the boundary layer will be turbulent when the free-stream plasma is also turbulent, the hydrodynamic turbulence directly affects the transport mechanism of the ionized species to the probe surface. Of course, a situation is often encountered wherein the free-stream plasma is steady and only the boundary layer is turbulent. Also, at times, the ionization takes place within the turbulent boundary layer, and the relationship between the current and potential measured at the surface and the turbulent properties of the electrons within the boundary layer are sought.

Reasonable progress has been made toward relating the measured mean values of the current and potential of a surface probe to the mean electron concentrations across the boundary layer only for the case when the sheath is embedded within the laminar sublayer. For this problem, it is only necessary to determine the mean ambipolar (quasi-neutral) concentration profiles of the ionized species for the given turbulent boundary layer. The results can then be employed along with the laminar sheath solutions to construct the current-potential relationship in the same manner as for the laminar boundary layer discussed in Chapter III. For this purpose Chapkis (1969), formulated the governing equations for the quasi-neutral turbulent boundary layer, and numerically integrated them for flow past a flat plate and cone, and obtained various mean profiles including those of the ionized species concentration. Baum and Denison (1972) matched the laminar sheath solutions to the quasi-neutral solutions and obtained current-potential characteristics for turbulent boundary layers.

The continuity, momentum, and energy equations employed by Chapkis (1969), and Baum and Denison (1972), are the standard equations based on the mixing-length theory. Along with these equations, Chapkis (1969), and Baum and Denison (1972) employed the following conservation equation for the ionized species for the case of a chemically frozen, quasi-neutral boundary-layer flow.

$$\bar{\rho}\bar{u}\frac{\partial \bar{c}}{\partial x} + \bar{\rho}\bar{v}\frac{\partial \bar{c}}{\partial y} = \frac{\partial}{\partial y}\left[\frac{\mu}{Sc_i}\left(\frac{2}{1+\beta} + Sc_i\frac{\epsilon\bar{\rho}}{\mu}\right)\frac{\partial \bar{c}}{\partial y}\right] \tag{4.18}$$

In the above equation, $\bar{c}$ is the mean mass fraction of the ions normalized with respect to a reference point. It is therefore the same as the similarly normalized mass fraction of the electrons. Sc_i is the molecular ion Schmidt number, and $\beta = (m_e/m_i)^{1/2} \ll 1$. ϵ is the eddy diffusivity defined by

$$\overline{\rho V C} = -\rho\epsilon\frac{\partial \bar{c}}{\partial y} \tag{4.19}$$

where C and V are the fluctuating components of c and v, respectively.

In Chapkis (1969) and Baum and Denison (1972), the eddy diffusivities for the transport of momentum, heat, and the ionized species are assumed to be all the same, and the expression proposed by Lees and Chapkis (1969) was employed for ϵ. The governing equations, including Eq. (4.18), were then integrated numerically and the streamwise variations of the ionized-species profile for a flat plate and an axisymmetric cone were determined for a given upstream profile.

We shall close this portion of our discussion by investigating the validity of Eq. (4.18). We first begin with the following fundamental governing equations of all weakly ionized continuum plasma boundary layers (see

Chapter III and the Appendix), valid when $T_i = T_e = T$.

$$\rho \frac{\partial c_i}{\partial t} + \rho u \frac{\partial c_i}{\partial x} + \rho v \frac{\partial c_i}{\partial y} = \frac{\partial}{\partial y}\left[\frac{\mu}{Sc_i}\left(\frac{\partial c_i}{\partial y} - \frac{e}{kT}\hat{\epsilon} c_i\right)\right] \tag{4.20}$$

$$\rho \frac{\partial c_e}{\partial t} + \rho u \frac{\partial c_e}{\partial x} + \rho v \frac{\partial c_e}{\partial y} = \frac{\partial}{\partial y}\left[\frac{\mu}{Sc_e}\left(\frac{\partial c_e}{\partial y} + \frac{e}{kT}\hat{\epsilon} c_e\right)\right] \tag{4.21}$$

$$\frac{\partial \hat{\epsilon}}{\partial y} = \frac{e}{\sigma}(c_i - c_e) \tag{4.22}$$

In the above equations, c_i and c_e are the normalized instantaneous mass fractions of ions and electrons, respectively. $\hat{\epsilon}$ is the instantaneous y component of the electric field intensity and e and σ are the electron charge and the permittivity, respectively.

Now, if and only if $c_i = c_e = c$ for all x, y, and time t, then Eq. (4.22) is superfluous (see Chapter III), and Eqs. (4.20) and (4.21) can be combined to yield

$$\rho\left(\frac{\partial c}{\partial t} + u \frac{\partial c}{\partial x} + v \frac{\partial c}{\partial y}\right) = \frac{\partial}{\partial y}\left(\frac{\mu}{Sc_i} \frac{2}{1 + \beta} \frac{\partial c}{\partial y}\right) \tag{4.23}$$

Following the standard practice, we now express the variables in terms of the mean quantities, $(\bar{\ })$, and the fluctuating quantities U, V, and C, as

$$\begin{aligned} u &= \bar{u} + U \\ v &= \bar{v} + V \\ c &= \bar{c} + C \end{aligned} \tag{4.24}$$

Equation (4.23) then becomes Eq. (4.18) for steady-state conditions with the assumption of Eq. (4.19).

Hence Eq. (4.18) is correct within the accuracy of the eddy-diffusivity concept represented by Eq. (4.19) provided that $T_i = T_e$ and $c_i(x, y, t) = c_e(x, y, t)$. Considering the present status of turbulent boundary-layer theory, the eddy-diffusivity concept is probably as acceptable as any other available theory. Furthermore for Damkohler numbers sufficiently large to ensure electron-neutral temperature equilibration, the assumption of $T_i = T_e$ should be satisfactory. The necessary criteria for the equality $c_i(x, y, t) = c_e(x, y, t)$ as well as a few other pertinent statistical turbulence properties of continuum plasmas will be discussed in the following.

Marcisz and Chung (1972) carried out a statistical spectrum analysis of a turbulent continuum plasma. In this work the dynamic equations of two-point correlation functions were derived from the fundamental governing equations,[18] which are similar to Eqs. (4.20)–(4.22). The closure of these

[18] These equations are the same as Eqs. (4.20)–(4.22) but no boundary-layer approximations have been made.

equations was accomplished by assuming that the turbulence was sufficiently weak such that the correlation functions of order higher than two were negligibly small. (This closure, as well as all other closures, has its defects.) In order to make the analysis tractable Marcisz and Chung (1972) considered that all mean gradients were uniform. A spectral analysis of the dynamic equations was then carried out using the Deissler (1970) solution of the neutral gas turbulence field with uniform velocity gradient.

The assumption of a weak turbulence field implies a field with low-to-moderate-turbulence Reynolds number, $Re_t = \frac{u'l}{\nu}$, where u' and l are the characteristic fluctuation velocity and the scale of the turbulent field (of the order of the integral scale), respectively. Therefore the implication of Marcisz and Chung's analysis for high Reynolds number boundary-layer flows must be considered with caution.

Marcisz and Chung showed that, in a turbulent shear flow, $\bar{c}_i(\vec{x}) = \bar{c}_e(\vec{x})$ does not necessarily imply $\bar{c}_i(\vec{x}, t) = \bar{c}_e(\vec{x}, t)$. It was shown that for the electrons to fluctuate with the ions, the parameter λ_D/l, where l is the local scale of turbulence and not the characteristic length of the flow field, must be sufficiently small, in addition to satisfaction of the condition $\bar{c}_e = \bar{c}_i$. Note that a decrease in l through the breakup of eddies is equivalent to an increase in λ_D through reduction of the electron concentration. The particular magnitude of λ_D/l, which is necessary for this detailed quasi-neutrality to exist depends on the concentration gradient $\hat{b} = d\bar{c}_e/dY$.

Figure 4-12 shows the one-point electron transport and electron mean-

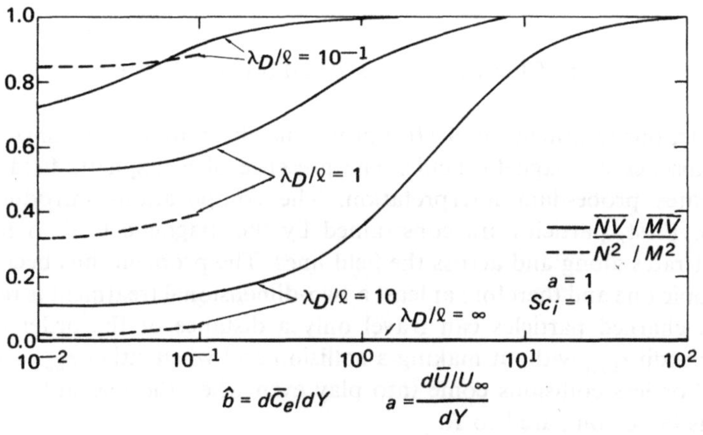

Fig. 4-12. One-point electron transport and electron-mean-square fluctuations relative to those of the ions as functions of λ_D/l and b. (From Marcisz and Chung, 1972.)

square fluctuation relative to those of the ions as functions of λ_D/l and $\hat{b}$. According to this figure, λ_D/l must be smaller than 10^{-1} (say $\lambda_D/l < 10^{-2}$) in order that the electrons fluctuate with the ions and give $\overline{NV}/\overline{MV} = 1$ and $\overline{N^2}/\overline{M^2} = 1$[19] for all parts of a shear flow field. $\lambda_D/l \leq 10^{-2}$, is then the criterion for the detailed quasi-neutrality of a low Re_t turbulent plasma.

A detailed quasi-neutrality must exist before the basic concept of Eq. (4.18) can be justified [see Eqs. (4.20)–(4.22)]. The particular value of λ_D/l required, however, will probably be different for high Re_t flows as compared to that predicted by Marcisz and Chung (1972) for low Re_t flows.

4.4.3 Transient Response (Laminar Plasmas)

Before ending this section we refer the reader back to Chapter II, Section 6 where collisionless probes operated in a transient mode were discussed. It was shown there that when collisionless probes are operated in the pulsed mode a current overshoot is obtained the magnitude of which can be used to determine the ion temperature. Cohen (1973) has predicted a similar current overshoot for continuum spherical probes in slightly ionized plasmas. The current overshoot is entirely an electron phenomenon and its magnitude is given by

$$I = -4\pi R D_{e0} N_{e0} \left\{ 1 + \chi_{p2} \left[1.355 + \frac{2}{3} \ln \frac{R}{\lambda_D} \right]^{-1} \right\} \tag{4.25}$$

where χ_{p2} is the final potential applied to the probe in a risetime t, which is of the order of electron-diffusion time. This is another example of the dissimilar transient response of electrons and ions.

4.5 Electric Probes in Magnetic Fields

So far, our treatment of electric probes has been limited to ionized gases in the absence of magnetic fields. The presence of a magnetic field further complicates probe-data interpretation. The complications introduced are twofold. First, particles are constrained by the magnetic field to move at different rates along and across the field lines. The problem thus becomes an anisotropic one and therefore at least a two-dimensional treatment is required. Second, charged particles can travel only a distance of the order of their Larmor radii $r_{L_{e,i}}$ without making a collision and when either r_{L_e} or r_{L_i} is of order R or less collisions come into play even when the relevant mean free path λ is large compared to R.

Chen (1965) has given a general treatment of probe theory in the presence

[19] M, N, and V denote the fluctuating portion of the ion number density, electron number density, and the y component of the velocity, respectively.

of magnetic fields and discussed several specific cases available in the literature prior to 1965. One of the more extensive mathematical treatments has been given by Bertotti (1961, 1962) for probes in strong B-fields. However, as pointed out by Chen (1965) and demonstrated by experiments, Bertotti's results are at best of limited applicability.

Treatment of probes in magnetic fields has continued to be sparse since Chen's review of 1965, despite the importance attached to applications to thermonuclear plasmas, MHD generators, and geophysical problems. There have been several important contributions, however. These include experiments on disc probes in weak fields by Sugawara (1966) and on cylindrical probes in strong fields by Brown *et al.* (1971). Theoretical studies have been carried out by Sanmartin (1970), who used a kinetic theory approach to probes in strong B-fields; Cohen (1969) and Niyogi and Cohen (1973), who considered continuum probes in magnetic fields, and Lam (1973) who introduced the magnetic sheath concept for probes in collisionless, fully ionized plasmas in a strong magnetic field.

First, we give a general sketch of the problem at hand. The current collection phenomena of probes in plasmas with magnetic fields can be divided into three broad categories. In order of increasing complexity, these are weak B-field case: $R \ll r_{L_{e,i}}$; moderate B-field case: $r_{L_e} < R < r_{L_i}$; and strong B-field case: $r_{L_{e,i}} < R$.

A more detailed classification of each of these regimes could be made by introducing various subclassifications depending on inequalities between λ, λ_D, and R such as were used in Chapter I to define the various regimes of probe operation in the absence of magnetic fields.

4.5.1 Weak *B*-Field Case

The simplest and most straightforward case arises when $r_{L_{e,i}} \left(= \dfrac{m_{e,i}V_\perp}{ZeB}\right)$, the gyro- or Larmor-radii of electrons and ions, are both larger than the probe radius and the Debye length. The probe *CV* characteristics are identical in this case to the collisionless or collisional zero-magnetic field solutions discussed in Chapters II and III. An example of the collisionless case is illustrated in the experiments of Chen *et al.* (1968) summarized in Fig. 4-13(a) and (b), where the ion-saturation current deviates from the Laframboise (1966) zero-field prediction only at low number densities and moderate B-fields ($n < 10^9\ \text{cm}^{-3}$; $B = 4$ kG).

4.5.2 Moderate *B*-Field Case

For $r_{L_e} \leq R$ anisotropies appear in the electron-transport coefficients and collision-like behavior becomes evident even if $\lambda < R$. Let us take the case first where $\lambda_D < R < \lambda$, which in the absence of a magnetic field would

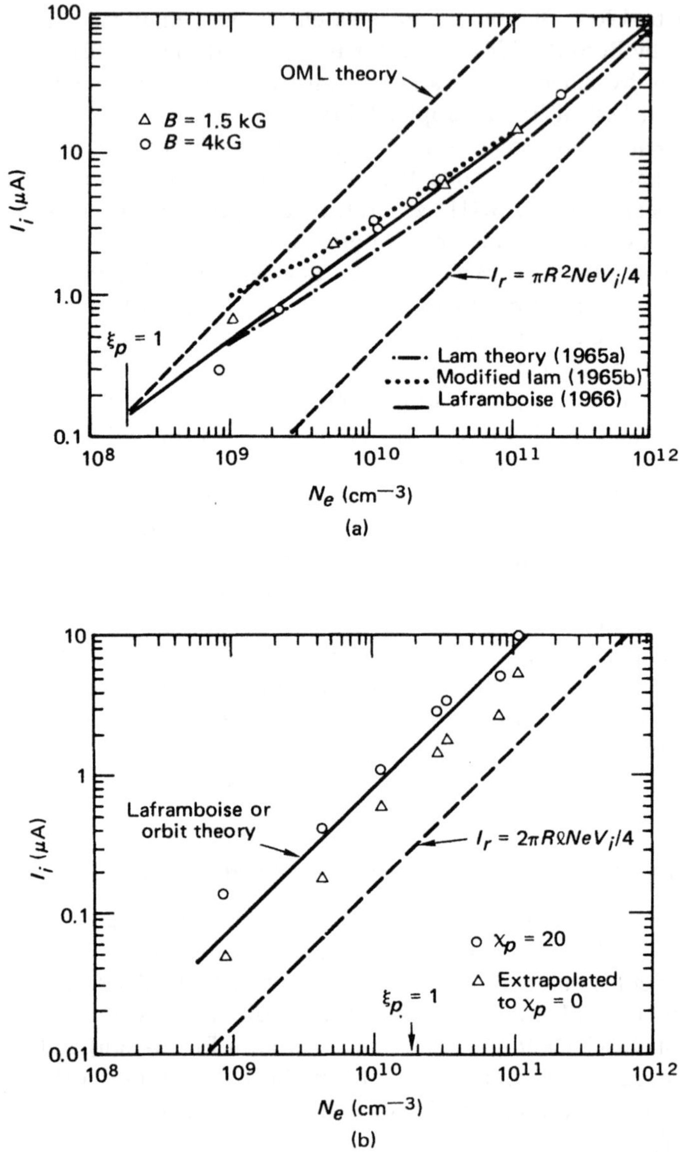

Fig. 4-13. (a) Measured values of ion-saturation current at $\chi_p = -20$ as a function of plasma density, as determined by microwaves, for two values of magnetic field. Probe radius 0.025 cm. (From Chen *et al.*, 1968.) (b) Measured values of ion-saturation current at $\chi_p = -20$ (circles), as a function of plasma density, as determined by microwaves, for $B = 4$ kG. Probe radius 0.0028 cm; probe length 0.19 cm. (From Chen *et al.*, 1968.)

correspond to the collisionless thin-sheath regime. If $\lambda_D \leq r_{L_{e,i}}$, the magnetic field remains approximately constant across the sheath, and the *thin*-sheath structure discussed in Chapter II is applicable here. In particular, for a negatively biased, absorbing surface the existence of the electric sheath requires that the Bohm criterion still be approximately satisfied, i.e., $v_i \geq \sqrt{\frac{kT_e}{m_i}}$ at the sheath edge. For $|\chi_p| \gg 1$, the repelled particles will be essentially Maxwellian and the magnitude of their Larmor radii will be unimportant provided $r_{L_{i,e}} < \lambda_D$. Here one can visualize a cylindrical tube defined by the magnetic lines of force intercepted by the probe. As with the zero-magnetic field case, ion collection at the probe is controlled by the Bohm condition, although its accurate value would be given by more sophisticated analyses, and therefore the effect of B on ion-saturation current is negligible. However, the electron collection can no longer be approximated by the random electron current in the plasma. This is because the diffusion of electrons across the field lines is reduced substantially, as can be seen from the expression for the diffusion coefficient

$$D_{\perp_e} = \frac{D_e}{1 + \omega_e^2 \tau_e^2} \tag{4.26}$$

where ω_e is the electron cyclotron frequency and τ_e is the mean collision time. For this case an approximate expression for the electron saturation current is given by Chen (1965), viz.

$$I_e = \tfrac{1}{4}(N_0 c_e A_p) \cdot \frac{4}{3}\frac{\lambda}{R}\sqrt{\frac{D_{\perp_e}}{D_e}} \tag{4.27}$$

where c_e is the electron thermal speed.

Because $\omega\tau$ for electrons is typically above 100, $D_\perp$ is reduced significantly even for weak fields. For ions, $\omega_i^2\tau_i^2$ is at least 2000 times smaller than for electrons. Thus $D_{\perp i}$ is affected only for very large values of B. The resulting anistropy for the electron flux gives rise to three very interesting phenomena.

1. Electron-saturation current is decreased below its value in the absence of a field. Experiments by Sugawara (1966) have demonstrated this. Sanmartin (1970) later predicted the decrease using a kinetic theory approach for a fully ionized collisionless gas. His results for the case of a spherical probe in a moderate magnetic field are shown in Fig. 4-14.

2. In the vicinity of the plasma potential, near the probe surface, an *overshoot* of potential is predicted. The physical reason for the overshoot is as follows. Electron flow across the B lines is strongly inhibited as described above, therefore, from mass conservation, the electron flux along B remains constant over long distances, although in a real case collisions would tend to

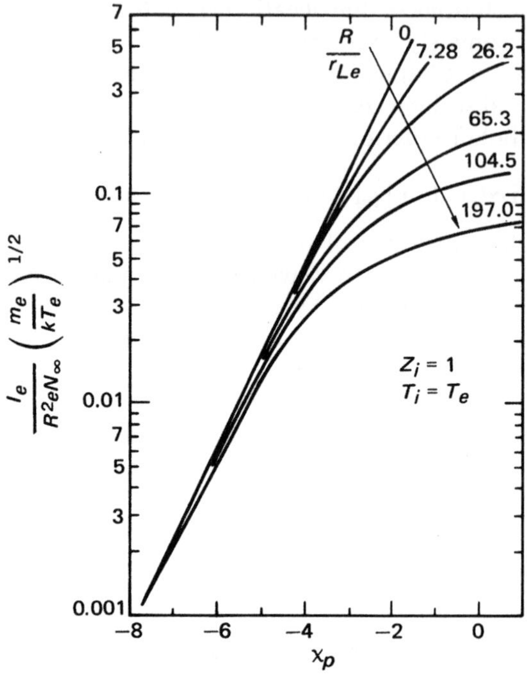

Fig. 4-14. Electron current as a function of χ_p and the ratio of electron cyclotron radius to probe radius. (From Sanmartin, 1970).

reduce this flux. For a perfectly absorbing probe there are no outgoing electrons from the probe surface so the electron density at the probe surface has to be very low. Because the magnetic field restricts the electron motion to essentially a unidimensional one, a potential overshoot is built up to meet the conditions of the low electron density near the probe. When the probe potential is increased or decreased from plasma potential, the overshoot disappears. Sanmartin's (1970) calculations exhibit this behavior for a fully ionized, low-density plasma in a moderate B-field. As a consequence of the overshoot, the usual sharp knee in the CV characteristics of the probe at space potential becomes blurred (see Fig. 4-14).

3. The variation in the cross section of the probe along B becomes unimportant for electron collection because of the channeling effect of the lines of force. Therefore the results of a spherical probe analysis can be applied to a thin disk perpendicular to B, and those of a cylinder to a two-dimensional strip.

As for ion-saturation current to a spherical probe, Sanmartin's results show slight distortion of the spherical symmetry, but little change in $I_{i(\mathrm{sat})}$, as one would expect, provided $r_{L_i} > R$. Furthermore, $n_e \approx \exp(\chi_p)$. Unfortunately Sanmartin was unable to demonstrate electron saturation because his numerical results were valid for $\chi_p \leq 0.5$ only.

In experiments conducted in a weakly ionized neon discharge, Sugawara (1966) obtained both electron- and ion-saturation conditions for disk probes under moderate B-field conditions, $B = 3000$ gauss. He found a linear slope for the quantity $d(\ln j_e)/dx_p$ from which he concluded that it was possible to determine the electron temperature. This was also found theoretically by Sanmartin (1970).

However, for strong B-fields, in tenuous plasmas ($\lambda > R$), it is possible to have nonequilibrium in the translational temperature with $T_\perp \neq T_\parallel$. In such a case, one would expect that the CV slope would yield $T_\parallel$ rather than T or $T_\perp$. An exact analysis for this case is as yet unavailable. It should also be noted that in strong B-fields, anomalous diffusion is a common phenomenon and $D_\perp$ is much larger than the value given by Eq. (4.28). This was not considered by Sanmartin (1970), but Sugawara (1966) consistently measured *higher* values of $j_{e_\perp}$ than he predicted from a simplified analysis when he used Eq. (4.26) for $D_\perp$.

An interesting experiment exhibiting the channeling effect of the magnetic field lines was carried out by Brown *et al.* (1971). They measured the ion current to cylindrical probes transverse to a discharge streaming along B-lines. A comparison was made between the number density measured by microwave interferometry $N_{\mu w}$ and the number density N_L using the Laframboise theory for cylindrical probes. The conditions of the experiment were $0.4 < R/\lambda_D < 4.0$; $\lambda \gg R$, and r_{L_i}/R of O(1), the moderate field case. They found that the ratio $N_L/N_{\mu w}$ as a function of magnetic field strength could be correlated by the simple expression.

$$\frac{N_L}{N_{\mu w}} = \frac{2RB + a}{2\pi RB + a} \tag{4.28}$$

where a is an experimentally determined constant. The rationale for this correlation is as follows. In the absence of a magnetic field $N_L = N_{\mu w}$ as it should. At the other extreme, for large B, the channeling effect of the field lines on the streaming plasma reduces the effective collection area of the probe to $2R$ per unit length whereas for vanishing B-fields, this area is $2\pi R$ per unit length. Therefore, in the limit of large B, $N_L/N_{\mu w} \rightarrow 1/\pi$. The data obtained by Brown *et al.* (1971) together with the correlation given by Eq. (4.28) are shown in Fig. 4-15. The success of this correlation which is very much in the same spirit as the interpolation formulas used in Chapter II for probes in the transition regime suggests that similar correlation formulas might be appropriate for other probe problems involving magnetic fields.

The continuum case of an electric probe in a strong, uniform B-field was

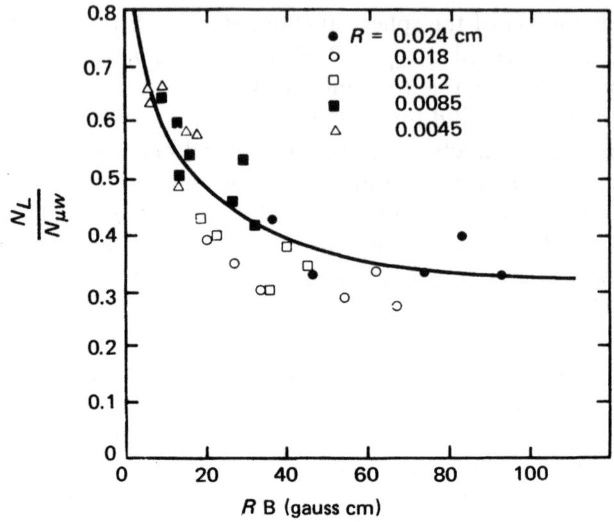

Fig. 4-15. The ratio $N_L/N_{\mu w}$ versus product RB (gauss · cm). The solid curve is a plot of Eq. (4.28) with a = 36 gauss · cm. (From Brown *et al.*, 1971.)

investigated by Cohen (1969) and Niyogi and Cohen (1973) for spherical and elliptic geometries in weakly ionized plasmas. Niyogi and Cohen (1973) obtained a numerical solution for the governing continuum probe equation in the limit $\lambda \ll \lambda_D \leq r_{L_e} \ll R$. The diffusion and mobility tensors were assumed constant. Most of the phenomena observed in the strongly ionized, rarefied plasma case above, were found to exist for continuum probes as well. For example, (1) an overshoot is observed in the potential, both along as well as normal to the B-field direction [see Fig. 4-16(a),(b)]; (2) there is a substantial decrease in $j_{e(\mathrm{sat})}$ with an increase in the magnitude of the B-field and an increase in probe size; (3) there appears a smoothing of the sharp knee in the CV characteristic at space potential; (d) approximate electron- and ion-current saturation are observed even in strong B-fields ($\omega_e\tau_e \geq 4$) as long as the ratio $R/\lambda_D \gg 1$. Niyogi and Cohen also show that[20]

$$I_{i,e\ \mathrm{sat}} \sim \left[1 + \left(\frac{T_e}{T_i}\right)^{\pm 1}\right]$$

exactly as with the zero-magnetic field case [Cohen (1969)]. Expressions are obtained for $I_{i,e(\mathrm{sat})}$ as functions of T_e/T_i, R/λ_D, and the semiaxis of the ellipsoid. For weak fields (e.g., $\omega_e\tau_e \leq 2$) where the plasma potential can be determined unambiguously, a simple expression is also obtained for the CV slope $d(ln\ I)/d(\chi_p)$.

[20] Positive sign for ion collection and vice versa.

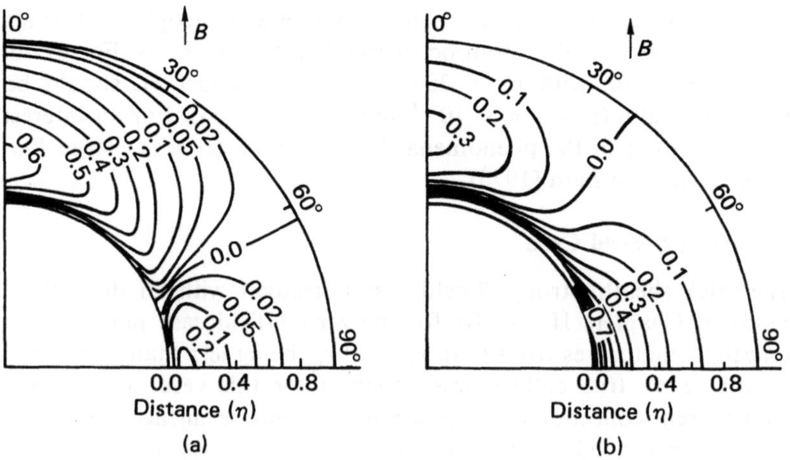

Fig. 4-16. Equipotential surfaces about a spherical probe: $R/\lambda_D = 100$; $T_e/T_i = 1$, $\omega_i^2\tau_{in}^2 = 0.01$, $\omega_e^2\tau_{en}^2 = 2.60$ (a) $\chi_p = 0$; (b) $\chi_p = -1$. θ is the polar angle measured from the magnetic axis and $\eta = 1 - \dfrac{R}{r}$.

4.5.3 Induced *B*-Field Effects

Up to this point we have limited our discussion to uniform, applied field B_0 with no induced field effects. When significant currents flow in the plasma, induced effects are given by Ampere's law,

$$\nabla \times \vec{B} = \mu_0 \vec{j} \tag{4.29}$$

where $\vec{j} = \sum_s Z_s e n_s V_s$.

If we use the Ohm's law expression for j, and assume E of order $(\vec{v} \times \vec{B})$, then Eq. (4.29) can be written in a more familiar form

$$\tilde{\nabla} \times \tilde{\vec{B}} = Rm\,\vec{j} \tag{4.30}$$

where $Rm = \mu_0 \sum_c u_\infty L$ is the magnetic Reynolds number, μ_0 is the permeability, σ_c is the electrical conductivity, u_∞ a reference velocity, and L a reference length for the plasma. When $Rm \gg 1$, small currents will induce large nonuniformities in the B-field. These nonuniformities will be restricted to an inviscid magnetic "boundary layer" of the order $Rm^{-1/2}$ adjacent to the probe surface, quite analogous to viscous boundary layers, which are of order $Re^{-1/2}$ ($Re = \rho u_\infty L/\mu$). One then would need to employ Ampere's law in addition to the flow conservation and Poisson equations discussed in Chapter III, to properly analyze the CV characteristics of an electric probe in a magnetic field. Needless to say, this will significantly complicate the problem. Situations in which $Rm \gg 1$ are fortunately limited to geophysical or astrophysical phenomena and seldom occur in laboratory plasmas. For example, to obtain $Rm = 10$, with $u_\infty = 300$ m/sec, the product σL should be of order 2×10^4 mho ($\mu_0 = 4\pi \times 10^{-7}$ volt · sec/amp · m). For an interesting discussion of some of the phenomena that occur for probes at large Rm the reader is referred to Lam (1973).

4.5.4 Strong *B*-Field Case

The thick sheath, strong *B*-field case is fraught with all the difficulties discussed in Chapters II and III for the zero *B*-field case plus the double complication that arises from both r_{L_e} and r_{L_i} being less than R. Obviously, there will be no true collisionless situation for this case, and unlike the electron-current-collection case, diffusion to the probe surface from the sides of the tube determined by the magnetic lines of force may become important during ion collection. The continuum, strong *B*-field computations of Niyogi and Cohen (1973) show the expected trend of a decrease in $I_{i(\text{sat})}$ as the field strength is increased. For low-density plasmas, the Sanmartin type solutions (Sanmartin 1970) will be especially complicated in that the potential overshoot may now appear in the sheath and the magnetic field may not remain constant through it. No solutions are available as yet for this case.

It is clear from the above considerations that the use of electrostatic probes in plasmas with magnetic fields should be limited, if possible, to situations that correspond to the weak *B*-field case where the *B*-field has negligible effect on the *CV* characteristics. For moderate *B*-field, the use of simple models and interpolation formulas such as that of Brown *et al.* (1971) may be the most efficacious method for interpreting probe data. Very little of practical use is known concerning probe response in strong *B*-fields.

References

Bailey, P. B. and Touryan, K. J. (1973), *AIAA J.*, **11**, 1225.

Barad, M. S. and Cohen, I. M. (1973), Univ. of Pennsylvania Rep. 2281/10, Philadelphia, Pennsylvania.

Barnes, B. T. and Eros, S. (1950), *J. Appl. Phys.*, **21**, 1275.

Baum, E. and Denison, M. R. (1972), TRW Rep. 18586-6031-R0-00, TRW Systems, Redondo Beach, California.

Berman, C. H. and Lam, S. H. (1968), AIAA Pap. No. 68-725, AIAA Fluid and Plasma Dynamics Conference, Los Angeles, California, June 1968.

Bertotti, B. (1961), *Phys. Fluids*, **4**, 1047.

Bertotti, B. (1962), *Phys. Fluids*, **5**, 1010.

Bienkowski, G. and Kalnavarns, J. (1968), AIAA Pap. No. 68-167, AIAA 6th Aerospace Sciences Meeting, New York, January 1969.

Brown, I. G., Compher, A. B. and Kunkel, W. B. (1971), *Phys. Fluids*, **14**, 1377.
Burke, A. F. (1967), CAL Rep. AN-2101-Y-1, Cornell Aeronautical Laboratory, Buffalo, New York.
Chang, K. W. and Bienkowski, G. (1970), *Phys. Fluids*, **13**, 902.
Chapkis, R. L. (1969), TRW Rep. 06488-6312-R3-00, TRW Systems, Redondo Beach, California.
Chen, F. F. (1965), *Plasma Diagnostic Techniques*, Academic Press, New York.
Chen, F. F., Etievant, C., and Mosher, D. (1968), *Phys. Fluids*, **11**, 811.
Cohen, I. M. (1969), *Phys. Fluids*, **12**, 361.
Cohen, I. M. (1973), *Phys. Fluids*, **16**, 700.
Crawford, F. W. (1963), *J. Appl. Phys.*, **34**, 1897.
Deissler, R. G. (1970), *Phys. Fluids*, **13**, 1868.
Demetriades, A. and Doughman, E. L. (1966), *AIAA J.*, **4**, 451.
Garscadden, A. and Emeleus, K. G. (1962), *Proc. Phys. Soc.*, **79**, Pt. 3, 535.
Hayes, D. T., Herskovitz, S. B., Lennon, J. F., and Poirier, J. L. (1974). *Spacecraft and Rockets* **11**, 387.
Hirao, K. and Oyama, K. (1972), *J. Geomag. Geoelectr.*, **24**, 415.
Johnson, D. H. and Touryan, K. J. (1967), Sandia Rep. SC-RR-67-289, Sandia Laboratories, Albuquerque, New Mexico.
Jou, W. H. and Cheng, S. I. (1971), *Phys. Fluids*, **14**, 2144.
Koopman, D. W. (1971), *Phys. Fluids*, **14**, 1707.
Laframboise, J. G. (1966), Univ. of Toronto Institute of Aerospace Studies Rep. 100.
Lam, S. H. (1973), in *Dynamics of Ionized Gases*, IUTAM Int. Symp. Dynamics of Ionized Gases held in Kyoto, Japan, 1971, University of Tokyo Press, Tokyo, Japan, pp. 193.
Langmuir, I. and Mott-Smith, H. M. (1926), *Phys. Rev.*, **28**, 727.
Lees, L. and Chapkis, R. L. (1969), *AIAA J.*, **7**, 671.
Luzzi, T. and Jenkins, R. (1971), *AIAA J.*, **9**, 2411.
Marcisz, T. J. and Chung, P. M. (1972), AIAA Pap. 72-107, 10th Aerospace Sciences Meeting, San Diego, California.
Medicus, G. (1961), 5th Int. Conf. Ionization Phenomena in Gases, **2**, 1397.
Nishida, M. (1972), *Phys. Fluids*, **15**, 596.
Niyogi, K. and Cohen, I. M. (1973), *Phys. Fluids*, **16**, 69.
Peterson, E. W. (1968), Univ. of California Aerospace Science Rep. AS-68-10, University of California, Berkeley, California.
Peterson, E. W. and Talbot, L. (1970), *AIAA J.*, **8**, 1391.
Sajben, M. (1970), *AIAA J.*, **8**, 400.
Seemann, G. R. and Thornton, J. A. (1969), AIAA Pap. 69-700, AIAA 2nd Fluid and Plasma Dynamics Conference, San Francisco, California, June 1969.
Sanmartin, J. R. (1970), *Phys. Fluids*, **13**, 103.
Shelton, G. (1970), Sandia Rep. SC-RR-70-331, Sandia Laboratories, Albuquerque, New Mexico.
Shelton, G., Touryan, K. J., and Johnson, D. H. (1968), *Phys. Fluids*, **11**, 2773.
Starner, K. E. (1966), *AIAA J.*, **4**, 1685.
Starner, K. E. (1969), *AIAA J.*, **7**, 2357.
Su, C. H. and Sonin, A. A. (1967), *Phys. Fluids*, **10**, 124.

Sugawara, M. (1966), *Phys. Fluids*, **9**, 797.
Sugawara, M. and Hatta, Y. (1963), Institute of Plasma Physics Rep. IPPJ-4, Nagoya University, Nagoya, Japan.
Sugimoto, S. and Nishida, M. (1972), Kyoto University Rep. C.P. 35, Kyoto University, Kyoto, Japan.
Sutton, G. W. (1969), *AIAA J.*, **7**, 193.
Swift, J. D. and Schwar, M. J. R. (1971), *Electric Probes for Plasma Diagnostics*, Iliffe Books, Ltd., London.
Touryan, K. J. and Chung, P. M. (1971), *AIAA J.*, **9**, 365.
Waymouth, J. F. (1959), *J. Appl. Phys.*, **30**, 1404.

APPENDIX

General Theory

Introduction

The Boltzmann equation, augmented by the appropriate Poisson equation, is the logical starting point of the present discussion of partially ionized gases.

Because of the long-range intermolecular forces that exist between ionized species, the closure argument for the Boltzmann equation based on two-particle interactions and molecular chaos has been questioned in regard to its applicability to highly ionized, high-energy plasmas. Several attempts have been made to improve the Boltzmann equation by employing the next higher order equation in the BBGKY hierarchy of the Liouville equation. Such an attempt for instance has led to the development of the Lenard–Balescu equation [see, e.g., Balescu (1960) and Lenard (1960)]. The present discussion, however, is concerned with relatively weakly ionized, low-energy plasmas, and therefore the Boltzmann equation is considered to be sufficiently accurate.

In what follows the Boltzmann equation will first be nondimensionalized so that the natural division of electric probe operation into the various regimes mentioned in Chapter I can be demonstrated. Also, from a study of the order of magnitudes of the various nondimensional terms, the basic mathematical problems confronted in these regimes can be foreseen.

Because one of the major subjects of this volume pertains to the continuum regime, the partial solution of the Boltzmann equation leading to the governing equations of a continuum plasma will be given in some detail. The emphasis in the derivation of the continuum equations will be on clarifying the logic and assumptions underlying these equations so that intelligent use of the continuum equations can be made.

A.1 Nondimensionalization of Boltzmann and Poisson Equations

We shall consider, for the purpose of discussion in this Appendix, a gas mixture consisting of three gaseous species. They are the electrons, positive ions, and the neutral species. These species will be signified by the three letters e, i, and a, respectively, which will be used as either subscripts or superscripts as needed. Often, ν will be employed to denote either e, i, or a. The superscript (0) or no superscript will be employed to signify the gaseous mixture. The absolute velocity vectors of the electrons, ions, and the neutral particles will be represented by $\vec{\xi}^e = \xi^e_\alpha$, $\vec{\xi}^i = \xi^i_\alpha$ and $\vec{\xi}^a = \xi^\alpha_a$, respectively, whereas their velocity vectors relative to the mass averaged velocity of the mixture, $\vec{u} = u_\alpha$, will be denoted by $\vec{C}^e = C^e_\alpha$, $\vec{C}^i = C^i_\alpha$, and $\vec{C}^a = C^a_\alpha$, respectively.

The Greek subscripts α, β, and γ will be reserved exclusively for Cartesian tensor indices.

Much of what immediately follows is drawn from Jeans (1925), Chapman and Cowling (1960), Sutton and Sherman (1965), and Burke (1967).

The following set of Boltzmann equations and the Poisson equation constitute the fundamental equations governing the entire problem at hand.

$$\frac{\partial f_e}{\partial t} + \xi^e_\alpha \frac{\partial f_e}{\partial x_\alpha} - \frac{e}{m^e} E_\alpha \frac{\partial f_e}{\partial \xi^e_\alpha} = \sum_{\nu = e,i,a} \int\int\int (f'_e f'_\nu - f_e f_\nu) g b db d\bar{\epsilon} d\vec{\xi}^\nu \quad \text{(A.1)}$$

$$\frac{\partial f_i}{\partial t} + \xi^i_\alpha \frac{\partial f_i}{\partial x_\alpha} + \frac{e}{m^i} E_\alpha \frac{\partial f_i}{\partial \xi^i_\alpha} = \sum_{\nu = e,i,a} \int\int\int (f'_i f'_\nu - f_i f_\nu) g b db d\bar{\epsilon} d\vec{\xi}^\nu \quad \text{(A.2)}$$

$$\frac{\partial f_a}{\partial t} + \xi^a_\alpha \frac{\partial f_a}{\partial x_\alpha} = \sum_{\nu = e,i,a} \int\int\int (f'_a f'_\nu - f_a f_\nu) g b db d\bar{\epsilon} d\vec{\xi}^\nu \quad \text{(A.3)}$$

$$\frac{\partial E_\alpha}{\partial x_\alpha} = \frac{e}{\sigma} \left(Z \int f_i d\vec{\xi}^i - \int f_e d\vec{\xi}^e \right) \quad \text{(A.4)}$$

In the above equations $f_\nu(t, \vec{x}, \vec{\xi}^\nu)$ is the distribution function, which is normalized such that it gives the particle-number density, n^ν, when integrated over the velocity space. E_α, m, and e denote the electric field intensity, particle mass, and the charge of an electron, respectively. Z and σ are the number of charges in the positive ions and the permittivity, respectively. Also, g, b, and $\bar{\epsilon}$ denote the usual relative speed between the colliding particles prior to the collision, impact parameter, and the impact angle, respectively. Finally, t and x_α are the time and the position vector, respectively, and $(\ )'$ denotes the value following the collision.

Now, we begin the nondimensionalization of the equations by first choosing the appropriate characteristic lengths and the velocity. These quantities as well as other characteristic reference quantities must be chosen such that the normalization of the terms of Eqs. (A.1)–(A.4) with respect to these quantities should render each of them order one at the most. There are three characteristic lengths associated with the problem. First is the characteristic length of the flow system L, which defines the region of the plasma flow. The other two are the characteristic lengths, which represent the range of the intermolecular forces and the range of the electrostatic force, respectively. If we let Q_∞ denote a reference-collision cross section, where ∞ denotes a suitable reference point, then the range of the intermolecular force is of the order of $(Q_\infty)^{1/2}$ where $(Q_\infty)^{1/2}$ is chosen as the characteristic length with which to nondimensionalize the impact parameter b. It is well known (e.g., Sutton and Sherman, 1965) that the range through which an electrostatic force can penetrate before being neutralized in an undisturbed plasma is of the order of the Debye shielding length λ_D, which is defined as $[\sigma kT^e/(e^2n^e)]^{1/2}$ where k is Boltzmann's constant. $\lambda_{D\infty}$ is thus chosen as the characteristic length with which to normalize the electric field intensities. It is noted here that, in the case of the collisions between the charged particles, the characteristic ranges of the intermolecular and electrostatic forces are the same.

All velocities in Eqs. (A.1)–(A.4) are molecular velocities. The average molecular speed of a gas in equilibrium[21] is $[8kT/(\pi m)]^{1/2} = O(kT/m)^{1/2}$. Therefore the quantity $(kT^\nu_\infty/m^\nu)^{1/2}$ will be used as the characteristic velocity for the νth species.

The other reference quantities are the obvious ones such as n^ν_∞ and T_∞ representing the orders of magnitude of the species number densities and the temperature, respectively, within the flow field.

The following dimensionless quantities are now defined in terms of the above characteristic quantities.

$$\tilde{f}_\nu = f_\nu\left[\frac{1}{n^\nu_\infty}\left(\frac{kT_\infty}{m^\nu}\right)^{3/2}\right]$$

$$\tilde{E}_\alpha = E_\alpha\left(\frac{\sigma}{en^e_\infty\lambda_{D\infty}}\right), \lambda_{D\infty} = \left(\frac{\sigma kT_\infty}{e^2n_{e\infty}}\right)^{1/2}$$

$$\tilde{t} = t\left[\frac{1}{L}\left(\frac{kT_\infty}{m^\nu}\right)^{1/2}\right]$$

$$\tilde{x}_\alpha = x_\alpha/L, \quad \tilde{\xi}^e_\alpha = \xi^e_\alpha\left(\frac{m^e}{kT_\infty}\right)^{1/2}$$

$$\tilde{\xi}^i_\alpha = \xi^i_\alpha\left(\frac{m^i}{kT_\infty}\right)^{1/2}, \quad \tilde{\xi}^a_\alpha = \xi^a_\alpha\left(\frac{m^a}{kT_\infty}\right)^{1/2}$$

$$\tilde{g} = g\left(\frac{m^\nu}{kT_\infty}\right)^{1/2}, \quad d\tilde{\epsilon}\tilde{b}d\tilde{b} = bdbd\epsilon/Q^\nu_\infty \tag{A.5}$$

With the use of Eqs. (A.5), Eqs. (A.1)–(A.4) are nondimensionalized as

[21] For the case of a Maxwellian velocity distribution.

$$\frac{\partial \tilde{f}_\nu}{\partial \tilde{t}} + \tilde{\xi}^\nu_\alpha \frac{\partial \tilde{f}_\nu}{\partial \tilde{x}_\alpha} - \Delta^\nu \frac{L}{\lambda_{D\infty}} \tilde{E}_\alpha \frac{\partial \tilde{f}_\nu}{\partial \tilde{\xi}^\nu_\alpha}$$

$$= \sum_{\bar{\nu}=e,i,a} n^{\bar{\nu}}_\infty L Q^{\nu\bar{\nu}}_\infty \int\int\int (\tilde{f}'_\nu \tilde{f}'_{\bar{\nu}} - \tilde{f}_\nu \tilde{f}_{\bar{\nu}}) \tilde{g} \tilde{b} \, d\tilde{b} \, d\tilde{\epsilon} \, d\vec{\tilde{\xi}}^{\bar{\nu}} \tag{A.6}$$

$$\frac{\partial \tilde{E}_\alpha}{\partial \tilde{x}_\alpha} = \frac{L}{\lambda_{D\infty}} (Z \int \tilde{f}_i d\tilde{\xi}^i - \tilde{f}_e d\tilde{\xi}^e) = \frac{L}{\lambda_{D\infty}} (Z\tilde{n}^i - \tilde{n}^e) \tag{A.7}$$

where Eq. (A.6) represents the three equations for $\nu = e$, i, and a. Also, $\Delta^e = 1$, $\Delta^i = -1$, and $\Delta^a = 0$.

Now, the theoretical problem in the analysis of the response of electrostatic probes is to solve the above set of equations subject to the appropriate boundary conditions, which describe the particular probe configuration and flow conditions. Before we enter into discussions of the solutions of the equations, we shall discuss their general behavior apropos of the electrical characteristics of plasmas.

A.2 General Behavior of the Boltzmann Equation

Let us first define in the standard fashion the mean free path, $\lambda^{\nu\bar{\nu}}$, of a species ν, for collisions with the species $\bar{\nu}$ as

$$\lambda^{\nu\bar{\nu}} = \frac{1}{n^{\bar{\nu}} Q^{\nu\bar{\nu}}} \tag{A.8}$$

then we have

$$n^{\bar{\nu}}_\infty L Q^{\nu\bar{\nu}}_\infty = L/\lambda^{\nu\bar{\nu}}_\infty \tag{A.9}$$

The ratio $\lambda^{\nu\bar{\nu}}_\infty/L$ is the Knudsen number for the collisions of species ν with species $\bar{\nu}$. There is a set of nine Knudsen numbers in the equations represented by Eq. (A.6) for all possible combinations of ν and $\bar{\nu}$.

We see that the general behavior of the governing equations represented by Eqs. (A.6) and (A.7) is determined by the orders of magnitude of the smallest of the Knudsen numbers and the parameter $\lambda_{D\infty}/L$. These two parameters, which arise from the Boltzmann equations that provide the starting point for all electrostatic probe theories, are the most fundamental parameters, and they have been listed as Group 1 in Table 1 of Chapter I. The other groups of parameters shown in Table 1 arise from the boundary conditions and from the particular fashion in which the Boltzmann equations are reduced for particular domains of probe operation. These secondary groups of parameters are discussed in the subsequent chapters as they arise in the appropriate analyses.

In Section A.2 we shall discuss the most important limiting behaviors of Eqs. (A.6) and (A.7) as governed by the two fundamental parameters, the smallest of the Knudsen numbers and $\lambda_{D\infty}/L$. Hereafter we shall refer to the smallest of the Knudsen numbers simply as the Knudsen number Kn.

A.2.1 Large Knudsen Number Limit

We see from Eq. (A.9) that the collision terms of Eq. (A.6) are of the order L/λ_∞ $(= 1/Kn)$. Let us consider that $L/\lambda_\infty \ll 1$, and then study Eqs. (A.6) and (A.7) for the three different combinations of the parameters L/λ_∞ and $L/\lambda_{D\infty}$ represented in the list under domain A ($Kn \gg 1$) of Chapter I. When $L/\lambda_\infty \ll 1$, the only characteristic length of the flow system is the size of the probe. Hence in this case $L = R$.

The smallest-order terms of Eq. (A.6) are the collision terms for the cases 1 and 2, since $L/\lambda_{D\infty} \gg L/\lambda_\infty$. Hence the governing equations for these two cases is the Eq. (A.6), with the right-hand side of the equation set to zero, and the Poisson equation, Eq. (A.7). The Boltzmann equation with no collision terms used in conjunction with Poisson's equation is called the Boltzmann–Vlasov equation.

For case 1, we have $L/\lambda_{D\infty} \gg 1$. Hence we see in Eq. (A.7) that the quantity within the parentheses must be of order of magnitude smaller than order one. This means that for $L/\lambda_{D\infty} \gg 1$ the plasma cannot sustain an appreciable charge separation for most of the flow regions. Therefore the sheath will be limited to a thin region.

For case 2, however, $L/\lambda_{D\infty} \ll 1$. Equation (A.7) shows, for this case, that the quantity within the parentheses can be of order one. The plasma can, therefore, sustain a substantial charge separation in a large portion of the flow region, and the sheath is thick.

Finally, in the third case under A of the list given in Chapter I we have $L/\lambda_{D\infty} \ll L/\lambda_\infty$. Therefore $L/\lambda_{D\infty}$ approaches zero much faster than does L/λ_∞. Hence, in order to study the effect of $\tilde{E}_\alpha$ on the plasma, the collision terms must be retained in the Boltzmann equations even though the Knudsen number is very large. According to Eq. (A.7), the sheath is again very thick.

A clear understanding of the order of magnitude of $\tilde{E}_\alpha$ is quite important in the discussion of the behavior of the governing equations for small Knudsen numbers. Before we continue our exploration of the equations, therefore, let us consider the range of variation of $\tilde{E}_\alpha$ in a little more detail.

A.2.2 Order of Magnitude of $\tilde{E}_\alpha$

It was assumed, in nondimensionalizing the governing equations into the forms of Eqs. (A.6) and (A.7), that $\tilde{E}_\alpha$ is at most of order one, as it is with the other nondimensional functions in these equations. The reasons why the other dimensionless functions should be at most of order one are quite obvious from the manner in which they have been normalized. The fact that we used the Debye shielding length as the length scale for the normalization of E_α, because it is the characteristic range of the electric field intensity in an undisturbed plasma, is not a sufficient reason for $\tilde{E}_\alpha$ being at most of order one.

In order that we may be able to estimate the magnitude of $\tilde{E}_\alpha$ correctly for the various flow regions, let us first consider briefly the manner in which the electric field enters into a physical problem involving an electric probe.

An electric probe, which consists of a conducting surface is a sink for the charged particles because the surface is highly catalytic for the electron–ion recombination and is, in addition, highly absorbent for the electrons. If the probe surface is electrically grounded to an infinite sink such that it can continuously accept all the charged particles impinging upon it, then a much greater number of electrons will reach the probe surface per unit time than the ions because of the extremely small electron mass as compared to the ion particle mass. If the plasma is infinitely large as compared to the probe size, we see that the electron–ion-current ratio will be of the order of their inverse ratio, $\sqrt{(m^iT^e)/(m^eT^i)}$.

We can change this current ratio as well as the absolute values of the currents by applying an electric field, usually a negative field to retard the electrons, at the probe surface by the use of an external source such as a battery. The manner in which the plasma responds to the applied electric field, that is, the manner in which the electron and ion currents vary with respect to the applied field at the probe surface, depends on the various properties of the plasma such as the electron-number density and temperature.

Although the characteristic range of an electric field intensity is the Debye shielding length, the effect, for instance, of the electric field generated at the probe surface on the plasma is not limited to the Debye shielding length. First of all, one can force the electric field to penetrate much deeper into the plasma from the surface than the distance of a Debye shielding length simply by applying an arbitrarily large potential at the surface. That is, as we have seen elsewhere in this book, the sheath thickness can be increased arbitrarily by applying a large surface potential (provided that electrical breakdown of the plasma does not occur).

Furthermore, the electric field is generally not confined to the sheath. If a net current is collected by the probe, there must of necessity be a conduction current flowing in the ambient plasma, and associated with this current will be an electric field, unless the conductivity of the plasma is infinite. Hence $\tilde{E}_\alpha$ is actually not zero for regions of the plasma beyond a Debye shielding length from the probe, although three-dimensional effects rapidly reduce $\tilde{E}_\alpha$ to a very small value outside the sheath and its contiguous diffusion layer. With these ideas in mind, let us consider the possible orders of magnitude of $\tilde{E}_\alpha$.

As we see from Eq. (A.5), $\tilde{E}_\alpha$ is normalized with respect to $en^e_\infty\lambda_{D\infty}$. The quantity of $en^e\lambda_D$ represents the electric field intensity necessary to clear all the electrons out of the space within a Debye shielding length. This, therefore, is the maximum order of magnitude of the electric field intensity which can exist locally in the absence of phenomena such as surface emission and plasma breakdown. Therefore $\tilde{E}_\alpha$, which is equal to $\sigma E_\alpha/(en_\infty^e \lambda_{D\infty})$, can become at most of order one near the surface.

If a potential of order much greater than this maximum field intensity times a Debye shielding length is applied at the surface, the electric field will penetrate further into the plasma beyond the Debye shielding length, clearing out the electrons. Therefore the sheath thickens until the applied potential is accommodated. $\tilde{E}_\alpha$, however, still remains of order one.

The region outside the sheath where the electron- and ion-number densities are the same to first order is defined as the ambipolar region, or, more precisely, the quasi-neutral region. One can readily show from the electron and the ion equations represented by Eq. (A.6) that $\tilde{E}_\alpha$ must be of $O(\lambda_{D\infty}/L)$ at most in order that the electron and ion number densities be the same to first order in the quasi-neutral region.

Having argued that $\tilde{E}_\alpha$ should be at most of order one in the sheath region and that $\tilde{E}_\alpha$ should be at most of order $\lambda_{D\infty}/L$ in the quasi-neutral region, let us now return to the discussion of the behavior of Eqs. (A.6) and (A.7).

A.2.3 Small Knudsen Number Limit

We consider the case wherein the smallest of the Knudsen numbers, referred to simply as the Knudsen number, is much smaller order of magnitude than order one. For the purpose of illustration, consider that λ^{ea}/L is that Knudsen number. Equation (A.6) then can be written for electrons as

$$\epsilon\left(\frac{\partial \tilde{f}_e}{\partial \tilde{t}} + \tilde{\xi}_\alpha^e \frac{\partial \tilde{f}_e}{\partial \tilde{x}_\alpha} - \frac{L}{\lambda_{D\infty}} \tilde{E}_\alpha \frac{\partial \tilde{f}_e}{\partial \tilde{\xi}_\alpha^e}\right) = \sum_{\bar{\nu}=e,i,a} \frac{(\lambda_\infty^{ea}/L)}{(\lambda_\infty^{e\bar{\nu}}/L)} \int\int\int (\tilde{f}_e' \tilde{f}_\nu' - \tilde{f}_e \tilde{f}_\nu) \tilde{g} \tilde{b}\, d\tilde{b}\, d\tilde{\epsilon}\, d\vec{\tilde{\xi}}^{\bar{\nu}} \tag{A.10}$$

where

$$\epsilon = \lambda_\infty^{ea}/L = \frac{1}{n_\infty^a L (Q^{ea})_\infty} \tag{A.11}$$

Other equations can be similarly transformed.

The general behavior of the plasma for $\epsilon \ll 1$ can be studied by exploring the behavior of Eq. (A.10).

Let us first consider the quasi-neutral region. As was mentioned in the preceding subsection, $\tilde{E}_\alpha$ is at most of order $\lambda_{D\infty}/L$. Hence, $\tilde{E}_\alpha L/\lambda_{D\infty}$ is at most of order one. We see from Eq. (A.10) that the continuum state, which is defined as the state of collision dominance, is established if $\epsilon \ll 1$. For $\epsilon \ll 1$, the right-hand side of Eq. (A.6) predominates over the left-hand side, and the distribution functions of all species are near-Maxwellian. As will be shown in the next section, such near-Maxwellian (near equilibrium) distributions lead to the construction of the continuum plasma equations.

Next, we consider the sheath region. There, as it was argued previously, $\tilde{E}_\alpha$ is at most of order one. Let us rewrite the left-hand side of Eq. (A.10) as

$$\epsilon\left(\frac{\partial \tilde{f}_e}{\partial \tilde{t}} + \xi_\alpha^e \frac{\partial \tilde{f}_e}{\partial \tilde{x}_\alpha}\right) - \delta \tilde{E}_\alpha \frac{\partial \tilde{f}_e}{\partial \xi_\alpha^e} \tag{A.12}$$

where

$$\delta = (\lambda^{ea}/\lambda_D)_\infty \tag{A.13}$$

Now we see that the continuum state is established within the sheath if and only if both ϵ and δ are of smaller orders of magnitude as compared to order one.

To summarize, then, the continuum analysis of the quasi-neutral region is valid if the Knudsen number is of smaller order of magnitude as compared to order one. In order to extend the continuum analysis to the sheath region, on the other hand, the Debye shielding length must be much greater than the mean free path in addition to the small Knudsen number.

Plasma flows for which both $\epsilon \ll 1$ and $\delta \ll 1$, so that continuum conditions are uniformly established for both quasi-neutral and sheath regions, are called "continuum plasma" flows.

The cases 1 and 2 under domain B ($Kn \ll 1$) of the list in Chapter I satisfy the continuum-plasma criteria. The parameter $L/\lambda_{D\infty}$ is much greater than one in case 1, whereas it is much smaller than one in case 2. Therefore using the same arguments we employed for the large Knudsen number, we see that the sheath will be very thin for case 1 whereas it will be thick for case 2.

When $\delta \gg 1$, there exist two singular regions near the surface according to Eq. (A.10). Within the first singular region, the right-hand side (the collision terms) of Eq. (A.10) is of order $1/\delta$ and is negligible to first order. The term with $\tilde{E}_\alpha$ is then balanced by the other terms on the left-hand side with $\tilde{x}_\alpha$ stretched by a suitable function of $\epsilon/\delta = \lambda_{D\infty}/L$ representing the thickness of this region. In the second singular region, all terms of Eq. (A.10) are of equal order with again $\tilde{x}_\alpha$ stretched by a suitable function of $1/\delta$ and ϵ representing its thickness. In the outer region (the quasi-neutral region), the continuum state exists, as was shown before, and the collision terms are of predominant order. After a solution is obtained for each of these regions, these three regions must then be asymptotically matched to obtain the complete description of the problem when $\epsilon \ll 1$ and $\delta \gg 1$. This is the case 3 of domain B ($Kn \ll 1$) of the list given in Chapter I.

We shall, in Sections A.3 and A.4, derive the continuum equations from Eqs. (A.6) and (A.7) for plasmas with $\epsilon \ll 1$ and $\delta \ll 1$. As was mentioned earlier, such continuum equations will be uniformly valid for the plasma probe domains given as the cases 1 and 2 under domain B ($Kn \ll 1$) of Chapter I and will be valid in the quasi-neutral region of case 3.

A.3 Distribution Function for Continuum Plasma

The distribution function for electrons, uniformly valid for continuum-plasma flows and valid for the quasi-neutral region when $\epsilon \ll 1$ and $\delta \gg 1$, will be obtained herein by solving Eq. (A.6).

Since we seek the solutions for small values of ϵ we first put Eq. (A.6) into the form of Eq. (A.10)

$$\epsilon\left(\frac{\partial \tilde{f}_\nu}{\partial \tilde{t}} + \tilde{\xi}_\alpha^\nu \frac{\partial \tilde{f}_\nu}{\partial \tilde{x}_\alpha}\right) - \Delta^\nu \delta \tilde{E}_\alpha \frac{\partial \tilde{f}_\nu}{\partial \tilde{\xi}_\alpha^\nu} = \sum_{\bar{\nu}=e,i,a} A^{\nu\bar{\nu}} \int\int\int (\tilde{f}_\nu' \tilde{f}_{\bar{\nu}}' - \tilde{f}_\nu \tilde{f}_{\bar{\nu}}) \tilde{g}\tilde{b}\,d\tilde{b}\,d\tilde{\epsilon}\,d\vec{\tilde{\xi}}^{\bar{\nu}} \tag{A.14}$$

The above equation represents the three equations for $\nu = e$, i, and a. Also, $\Delta^e = 1$, $\Delta^i = -1$, and $\Delta^a = 0$. Furthermore in Eq. (A.14)

$$\epsilon = \lambda_\infty^{ea}/L = 1/[n_\infty^a L(Q^{ea})_\infty], \quad \delta = \lambda_\infty^{ea}/\lambda_{D\infty}$$

$$A^{e\bar{\nu}} = \lambda_\infty^{ea}/\lambda_\infty^{e\bar{\nu}} = \frac{n_\infty^{\bar{\nu}}(Q^{e\bar{\nu}})_\infty}{n_\infty^a(Q^{ea})_\infty} \tag{A.15}$$

$$A^{i\bar{\nu}} = \frac{n_\infty^{\bar{\nu}}(Q^{i\bar{\nu}})_\infty}{n_\infty^a(Q^{ea})_\infty}, \quad A^{a\bar{\nu}} = \frac{n_\infty^{\bar{\nu}}(Q^{a\bar{\nu}})_\infty}{n_\infty^a(Q^{ea})_\infty}$$

The parameters $A^{e\bar{\nu}}$, $A^{i\bar{\nu}}$, and $A^{a\bar{\nu}}$ are at most of order one.

After the method of Enskog–Chapman expansion [see Hirschfelder *et al.* (1954)] we seek a solution using a series expansion in powers of ϵ and δ.

$$\tilde{f}_\nu = \tilde{f}_{\nu 0} + \epsilon\tilde{\varphi}_{\nu 1} + \delta\tilde{\psi}_{\nu 1} + O(\epsilon^2, \delta^2, \epsilon\delta) \tag{A.16}$$

In the following, the zeroth and the first-order solutions of the distribution functions will be derived in terms of the nonuniformities in the plasma properties and the electric field intensity. The Poisson equation, Eq. (A.7), which defines the value of the electric field intensity, is not needed for this purpose, and its consideration will therefore be postponed until later.

A.3.1 Zeroth-Order Solutions

Substitution of the three series represented by Eq. (A.16) into the appropriate equation of Eq. (A.14) generates sets of various order perturbation equations, from which the zeroth-order equations are as follows:

$$\sum_{\bar{\nu}=e,i,a} A^{e\bar{\nu}} \int\int\int (\tilde{f}_{e0}' \tilde{f}_{\bar{\nu}0}' - \tilde{f}_{e0} \tilde{f}_{\bar{\nu}0}) \tilde{g}\tilde{b}\,d\tilde{b}\,d\tilde{\epsilon}\,d\vec{\tilde{\xi}}^{\bar{\nu}} = 0 \tag{A.17}$$

$$\sum_{\bar{\nu}=e,i,a} A^{i\bar{\nu}} \int\int\int (\tilde{f}_{i0}' \tilde{f}_{\bar{\nu}0}' - \tilde{f}_{i0} \tilde{f}_{\bar{\nu}0}) \tilde{g}\tilde{b}\,d\tilde{b}\,d\tilde{\epsilon}\vec{\tilde{\xi}}^{\bar{\nu}} = 0 \tag{A.18}$$

$$\sum_{\bar{\nu}=e,i,a} A^{a\bar{\nu}} \int\int\int (\tilde{f}_{a0}' \tilde{f}_{\bar{\nu}0}' - \tilde{f}_{a0} \tilde{f}_{\bar{\nu}0}) \tilde{g}\tilde{b}\,d\tilde{b}\,d\tilde{\epsilon}\,d\vec{\tilde{\xi}}^{\bar{\nu}} = 0 \tag{A.19}$$

The above zeroth-order equations are identical to those generated for a mixture of electrically neutral gases. After the treatment of a set of two zeroth-order equations for two-component gas mixtures by Chapman and Cowling (1960), which led to the H-theorem for the gas mixtures, it can be readily shown that the solutions of Eqs. (A.17)–(A.19) are the following Maxwellian distributions. The solutions have been given here in dimensional form for clarity.

$$f_{e0} = n^e\left(\frac{m^e}{2\pi kT}\right)^{3/2} \exp -\left(\frac{m^e C_e^2}{2kT}\right) \tag{A.20}$$

$$f_{i0} = n^i\left(\frac{m^i}{2\pi kT}\right)^{3/2} \exp -\left(\frac{m^i C_i^2}{2kT}\right) \tag{A.21}$$

$$f_{a0} = n^a\left(\frac{m^a}{2\pi kT}\right)^{3/2} \exp -\left(\frac{m^a C_a^2}{2kT}\right) \tag{A.22}$$

where

$$\begin{aligned} C_e^2 &= C_\alpha^e C_\alpha^e \\ C_i^2 &= C_\alpha^i C_\alpha^i \\ C_a^2 &= C_\alpha^a C_\alpha^a \end{aligned} \tag{A.23}$$

The important point to notice here is that the exact solutions, Eqs. (A.20)–(A.22) are all Maxwellian distribution functions about the *same* temperature T. Therefore the continuum transport equations derived in the conventional manner from the first-order perturbations of the Maxwellian functions, Eqs. (A.20)–(A.22), will not allow an electron temperature that differs from the neutral gas temperature.

If, however, one makes use of the Lorentz approximation from the beginning, which is usually employed at a later stage to evaluate the collision integral, then the Enskog–Chapman expansion can be consistently carried out to derive the continuum equations which allow for different electron temperatures.

The Lorentz approximation ignores the thermal motion of the heavy particles, neutrals and ions, as compared to the thermal velocities of electrons. Implicit in this approximation are that the electron energy is conserved and that the heavy-gas distribution functions are unchanged through the collisions between the electrons and the heavy gas particles. This approximation is valid for all practical purposes because of the extremely small electron to heavy-gas particle-mass ratio.

With the Lorentz approximation, let us reconsider Eqs. (A.17)–(A.19). First, Eq. (A.17) can be rewritten in the following manner since the heavy gas distributions are unaffected by the collisions with the electrons

$$A^{ee}\int\int\int(\tilde{f}'_{e0}\tilde{F}'_{e0}-\tilde{f}_{e0}\tilde{F}_{e0})\tilde{g}\tilde{b}d\tilde{b}d\tilde{\epsilon}d\vec{\tilde{\xi}}^{e}+A^{ei}\int\int\int\tilde{f}'_{i0}(\tilde{f}_{e0}-\tilde{f}'_{e0})\tilde{g}\tilde{b}d\tilde{b}d\tilde{\epsilon}d\vec{\tilde{\xi}}^{i}$$
$$+\int\int\int\tilde{f}_{a0}(\tilde{f}'_{e0}-\tilde{f}_{e0})\tilde{g}\tilde{b}d\tilde{b}d\tilde{\epsilon}d\vec{\tilde{\xi}}^{a}=0 \qquad \text{(A.24)}$$

F_{e0} is the zeroth-order distribution function of electrons as is f_{e0}, but the different symbol is used to distinguish it from f_{e0} as the collision partner of f_{e0}. The solution of Eq. (A.24) is, irrespective of f_{i0} and f_{a0},

$$f_{e0}=n^{e}\left(\frac{m^{e}}{2\pi kT^{e}}\right)^{3/2}\exp-\left(\frac{m^{e}C_{e}^{2}}{2kT^{e}}\right) \qquad \text{(A.25)}$$

where the electron temperature T^{e} is independent of the heavy-gas temperature T.

Solution of the remaining equations, Eqs. (A.18) and (A.19), gives the same Maxwellian distribution functions, Eqs. (A.21) and (A.22).

A.3.2 First-Order Solution

The perturbation equations of first order in ϵ and δ, resulting from the substitution of the three equations represented by Eq. (A.16) into those represented by Eq. (A.14) are as follows.

For order ϵ

$$\frac{\partial\tilde{f}_{\nu0}}{\partial\tilde{t}}+\tilde{\xi}^{\nu}_{\alpha}\frac{\partial\tilde{f}_{\nu0}}{\partial\tilde{x}_{\alpha}}=\sum_{\bar{\nu}=e,i,a}A^{\nu\bar{\nu}}\int\int\int(\tilde{\varphi}'_{\nu1}\tilde{f}'_{\bar{\nu}0}-\tilde{\varphi}_{\nu1}\tilde{f}_{\bar{\nu}0})\tilde{g}\tilde{b}d\tilde{b}d\tilde{\epsilon}d\vec{\tilde{\xi}}^{\bar{\nu}}$$
$$+\sum_{\bar{\nu}=e,i,a}A^{\nu\bar{\nu}}\int\int\int(\tilde{f}'_{\nu0}\tilde{\varphi}'_{\bar{\nu}1}-\tilde{f}_{\nu0}\tilde{\varphi}_{\bar{\nu}1})\tilde{g}\tilde{b}d\tilde{b}d\tilde{\epsilon}d\vec{\tilde{\xi}}^{\bar{\nu}} \qquad \text{(A.26)}$$

and for order δ

$$-\Delta^{\nu}\tilde{E}_{\alpha}\frac{\partial\tilde{f}_{\nu0}}{\partial\tilde{\xi}^{\nu}_{\alpha}}=\sum_{\bar{\nu}=e,i,a}A^{\nu\bar{\nu}}\int\int\int(\tilde{\psi}'_{\nu1}\tilde{f}'_{\bar{\nu}0}-\tilde{\psi}_{\nu1}\tilde{f}_{\bar{\nu}0})\tilde{g}\tilde{b}d\tilde{b}d\tilde{\epsilon}d\vec{\tilde{\xi}}^{\bar{\nu}}$$
$$+\sum_{\bar{\nu}=e,i,a}A^{\nu\bar{\nu}}\int\int\int(\tilde{f}'_{\nu0}\tilde{\psi}'_{\bar{\nu}1}-\tilde{f}_{\nu0}\tilde{\psi}_{\bar{\nu}1})\tilde{g}\tilde{b}d\tilde{b}d\tilde{\epsilon}d\vec{\tilde{\xi}}^{\bar{\nu}} \qquad \text{(A.27)}$$

Equations (A.26) and (A.27) each represent the three equations for $\nu = e, i$, and a.

From the solutions of the above equations, the distribution functions for electrons, ions, and neutral gas particles, which are accurate through the first order in ϵ and δ, will be constructed by the use of the series represented by Eq. (A.16).

Since the orders of magnitude of the various terms and equations have been firmly established through the appropriate nondimensionalizations, we can now in the interest of increased clarity operate in the dimensional world without confusion in orders of magnitude. Also, for convenience, we shall solve Eqs. (A.26) and (A.27) together.

We first rewrite the three series represented by Eq. (A.16) as

$$\begin{aligned}\tilde{f}_e &= \tilde{f}_{e0} + \tilde{f}_{e1}\\ \tilde{f}_i &= \tilde{f}_{i0} + \tilde{f}_{i1}\\ \tilde{f}_a &= \tilde{f}_{a0} + \tilde{f}_{a1}\end{aligned} \tag{A.28}$$

where

$$\begin{aligned}\tilde{f}_{e1} &= \epsilon\tilde{\varphi}_{e1} + \delta\tilde{\psi}_{e1}\\ \tilde{f}_{i1} &= \epsilon\tilde{\varphi}_{i1} + \delta\tilde{\psi}_{i1}\\ \tilde{f}_{a1} &= \epsilon\tilde{\varphi}_{a1} + \delta\tilde{\psi}_{a1}\end{aligned} \tag{A.29}$$

and the expression, $0(\epsilon^2, \delta^2, \epsilon\delta)$, has been dropped from the series for simplicity.

We then return Eqs. (A.28) to the dimensional form by multiplying the three equations by $n^e\left(\frac{m^e}{kT^e}\right)^{3/2}$, $n^i\left(\frac{m^i}{kT}\right)^{3/2}$, and $n^a\left(\frac{m^a}{kT}\right)^{3/2}$, respectively, to obtain

$$f_e = f_{e0} + f_{e1} \tag{A.30}$$

$$f_i = f_{i0} + f_{i1} \tag{A.31}$$

$$f_a = f_{a0} + f_{a1} \tag{A.32}$$

Note that the order relations remain intact so that, for instance f_{e1} is of order ϵ or δ, whichever is larger, as compared to f_{e0}.

Next we multiply Eq. (A.26) by ϵ, and Eq. (A.27) by δ. Then, we add these two equations, for each ν, and when we restore the resulting equations to dimensional form, the following equations result.

$$\begin{aligned}\frac{\partial f_{e0}}{\partial t} + \xi_\alpha^e \frac{\partial f_{e0}}{\partial x_\alpha} - \frac{e}{m^e}E_\alpha \frac{\partial f_{e0}}{\partial \xi_\alpha^e} &= \sum_{\bar{\nu}=e,i,a} \iiint (f'_{e1}f'_{\bar{\nu}0} - f_{e1}f_{\bar{\nu}0})gbdbd\bar{\epsilon}d\bar{\xi}^{\bar{\nu}}\\ &\quad + \sum_{\bar{\nu}=e,i,a} \iiint (f'_{e0}f'_{\bar{\nu}1} - f_{e0}f_{\bar{\nu}1})gbdbd\bar{\epsilon}d\bar{\xi}^{\bar{\nu}}\end{aligned} \tag{A.33}$$

$$\begin{aligned}\frac{\partial f_{i0}}{\partial t} + \xi_\alpha^i \frac{\partial f_{i0}}{\partial x_\alpha} + \frac{e}{m^i}E_\alpha \frac{\partial f_{i0}}{\partial \xi_\alpha^i} &= \sum_{\bar{\nu}=e,i,a} \iiint (f'_{i1}f'_{\bar{\nu}0} - f_{i1}f_{\bar{\nu}0})gbdbd\bar{\epsilon}d\bar{\xi}^{\bar{\nu}}\\ &\quad + \sum_{\bar{\nu}=e,i,a} \iiint (f'_{i0}f'_{\bar{\nu}1} - f_{i0}f_{\bar{\nu}1})gbdbd\bar{\epsilon}d\bar{\xi}^{\bar{\nu}}\end{aligned} \tag{A.34}$$

$$\begin{aligned}\frac{\partial f_{a0}}{\partial t} + \xi_\alpha^a \frac{\partial f_{a0}}{\partial x_\alpha} &= \sum_{\bar{\nu}=e,i,a} \iiint (f'_{a1}f'_{\bar{\nu}0} - f_{a1}f_{\bar{\nu}0})gbdbd\bar{\epsilon}d\bar{\xi}^{\bar{\nu}}\\ &\quad + \sum_{\bar{\nu}=e,i,a} \iiint (f'_{a0}f'_{\bar{\nu}1} - f_{a0}f_{\bar{\nu}1})gbdbd\bar{\epsilon}d\bar{\xi}^{\bar{\nu}}\end{aligned} \tag{A.35}$$

The left-hand sides of the above equations are functions of the zeroth-order distribution functions given by Eq. (A.25), and Eqs. (A.21) and (A.22). Equations (A.33)–(A.35) can be solved for f_{e1}, f_{i1}, and f_{a1} in the standard manner (Jeans, 1925; Chapman and Cowling, 1960; and Sutton and Sherman, 1965).

In particular, these equations can be solved by following the method Sutton and Sherman (1965) described for the solution of their analogous equations. For instance, the present Eq. (A.33) for f_{e1} is of the same form as Eq. (*e*) for $f_{1\nabla}$ found in Section 5.7 of Sutton and Sherman when $\vec{B} = 0$, except that the present equation contains the electric field term on the left. In following this method, however, it should be remembered that the electron temperature is different from the heavy-gas temperature in the present formulation, whereas all temperatures were considered to be the same in Sutton and Sherman. This means that certain conservation relations and the relationships between the species concentrations, species temperatures, and partial pressures employed in the solution of the present equations are different from those employed in Sutton and Sherman (1965).

Now, focusing our attention to the electrons, the solution of Eq. (A.33) gives

$$f_{e1} = f_{e0}\left[-\frac{G_\alpha C_\alpha^e}{C_e(n^i Q^{ei} + n^a Q^{ea})} + \frac{H_{\alpha\beta}(C_\alpha^e C_\beta^e - \frac{1}{3}C_\gamma^e C_\gamma^e \delta_{\alpha\beta})}{C_e(n^i s^i + n^a s^a)}\right] \tag{A.36}$$

where

$$G_\alpha = \frac{1}{p^e}\frac{\partial p^e}{\partial x_\alpha} + \left(\frac{m^e C_\gamma^e C_\gamma^e}{2kT^e} - \frac{5}{2}\right)\frac{1}{T^e}\frac{\partial T^e}{\partial x_\alpha} + \frac{e}{kT^e}E_\alpha - \frac{m^e}{\rho k T^e}\frac{\partial P}{\partial x_\alpha}$$

$$H_{\alpha\beta} = \frac{m^e}{kT^e}\frac{\partial u_\alpha}{\partial x_\beta} \tag{A.37}$$

In the above equations, C_e is the electron speed $(C_\alpha^e C_\alpha^e)^{1/2}$. S^i and S^a are functions of C_e, which are not needed for the present discussion, and therefore are not defined here. P and p^ν are the total pressure of the gas mixture and the partial pressure of the νth species, respectively.

The expressions similar to (A.36) can be obtained readily from Eqs. (A.34) and (A.35) for the ions and neutrals.

A.3.3 Distribution Function of Electrons Accurate through First Order

With the zeroth and first-order distribution functions given by Eqs. (A.25) and (A.36), respectively, the electron-distribution function accurate through first order is constructed as the sum of the two according to Eq. (A.30).

The ion and neutral gas distribution functions valid through the first order can be found by the same method. The results are of the same form as that for f_e except they are in terms of the heavy-gas temperature T, and G_α for the neutrals does not include the electric field.

A.4 Electron-Diffusion and Energy-Transport Vectors for Continuum Plasmas

By use of the distribution functions derived in Section A.3, we shall obtain expressions for the electron-diffusion and electron-energy transport, which are valid for small but nonvanishing values of the Knudsen number ϵ.

A.4.1 Electron Diffusion

Electron diffusion is defined herein as the transport of electrons relative to the average velocity of the gas mixture. We therefore seek the expression for

$$m^e n^e \langle C_\alpha^e \rangle = m^e f_e C_\alpha^e d\vec{C}^e = m^e\left(\int f_{e0} C_\alpha^e d\vec{C}^e + \int f_{e1} C_\alpha^e d\vec{C}^e\right) \quad \text{(A.38)}$$

where Eq. (A.30), which is valid for small values of ϵ, has been employed. The symbol $\langle \quad \rangle$ will be employed henceforth to signify the ensemble average based on the species distribution function.

With the use of Eqs. (A.25) and (A.36), Eq. (A.38) becomes

$$\begin{aligned} m^e n^e \langle C_\alpha^e \rangle = & -m^e\left(\frac{1}{p^e}\frac{\partial p^e}{\partial x_\beta} + \frac{eE_\beta}{kT^e} - \frac{m^e}{\rho k T^e}\frac{\partial P}{\partial x_\beta}\right)\int f_{e0}\frac{C_\alpha^e C_\beta^e}{C_e(n^i Q^{ei} + n^a Q^{ea})} d\vec{C}^e \\ & - m^e \frac{1}{T^e}\frac{\partial T^e}{\partial x_\beta}\int f_{e0}\frac{C_\alpha^e C_\beta^e[(m^e C_e^2)/(2kT^e) - 5/2]}{C_e(n^i Q^{ei} + n^a Q^{ea})} d\vec{C}^e \\ & + \frac{(m^e)^2}{kT^e}\frac{\partial u_\gamma}{\partial x_\beta}\int f_{e0}\frac{(C_\gamma^e C_\beta^e - \frac{1}{3}C_e^2\delta_{\gamma\beta})C_\alpha^e}{C_e(n^i S^i + n^a S^a)} d\vec{C}^e \end{aligned} \quad \text{(A.39)}$$

In the last term of the above equation the integral, apart from the factor f_{e0}, is always anisotropic. Because f_{e0} is isotropic, this integral vanishes.

The analogous term which will subsequently appear in the electron energy transport relation will also be zero. Therefore, as stated earlier, the actual S^v are not needed for the present discussion.

With the last term having vanished, Eq. (A.39) shows that the electron diffusion is caused by the electron and total pressure gradients, the electric field intensity, and the electron temperature gradient, The second term, which is proportional to the electron temperature gradient, describes the contribution to electron diffusion called "thermal diffusion." This term should not be confused with a term arising from the $\partial p^e/\partial x_\beta$; contained in the first term of Eq. (A.39), which is also proportional to the electron-temperature gradient, because the electron pressure gradient term is usually split into a term involving the concentration gradient and one involving the electron temperature gradient. This term involving the electron temperature gradient is not the thermal diffusion term.

With regard to the thermal diffusion term, the second term of Eq. (A.39), the integral vanishes identically if the Maxwell inverse-fifth power law is used in the evaluation of the collision integrals. For the other intermolecular force laws, this integral is nonzero. It is, however, usually very small as compared to the integral in the first term. Therefore thermal diffusion can generally be neglected even for intermolecular force laws other than the Maxwell force law unless the electron temperature gradient is extremely large.

Next, we note that the term $\dfrac{m^e}{\rho k T^e}\dfrac{\partial p}{\partial x_\beta}$ in the factor multiplying the first integral on the right of Eq. (A.39) represents a part of the pressure diffusion. However, most of the pressure diffusion is accounted for by the partial pressure gradient term $\dfrac{\partial p^e}{\partial x_\beta}$. In fact $\dfrac{m^e}{\rho k T^e}\dfrac{\partial p}{\partial x_\beta}$ is negligibly small as compared to the partial pressure gradient term because m^e is extremely small, and it will be dropped henceforth. It should be noted that the dropping of this term does not imply the neglect of the pressure diffusion. On the contrary, the pressure diffusion is almost exactly taken into account by the partial-pressure-gradient term.

Equation (A.39) finally becomes

$$\langle C_\alpha^e \rangle = -D^e\left(\frac{1}{p^e}\frac{\partial p^e}{\partial x_\alpha} + \frac{eE_\alpha}{kT^e}\right) \tag{A.40}$$

where D^e is the diffusion coefficient of the electrons defined by

$$D^e = \frac{4\pi}{3}\frac{1}{n^e}\int_0^\infty f_{e0}\,\frac{C_e^3}{n^i Q^{ei} + n^a Q^{ea}}\,dC_e \tag{A.41}$$

In classical studies of ionized gases wherein electron diffusion is considered to result from the action of electric field alone, the electric conductivity or the electron mobility K^e is defined such that

$$\langle C_\alpha^e \rangle = K^e E_\alpha \tag{A.42}$$

By incorporating the above definition of electron mobility into Eq. (A.40) we obtain

$$\langle C_\alpha^e \rangle = -\left(D^e \frac{1}{p^e}\frac{\partial p^e}{\partial x_\alpha} + K^e E_\alpha\right) \tag{A.43}$$

where

$$\frac{D^e}{K^e} = \frac{kT^e}{e} \tag{A.44}$$

The above relationship between the diffusion coefficient and the mobility is referred to as the "Einstein" relationship.

Consider a particular situation wherein $T^e = T$ and the total pressure P

is constant in the direction of diffusion. Then, Eq. (A.43) is readily put into the form

$$m^e n^e \langle C_\alpha^e \rangle = \rho^e \langle C_\alpha^e \rangle = -\rho\left(D^e \frac{\partial N^e}{\partial x_\alpha} + K^e E_\alpha N^e \right) \quad \text{(A.45)}$$

where ρ^e and ρ are the partial density of electrons and the total density of the gas mixture, respectively. N^ν is the mass fraction and is defined by $\dfrac{n^\nu m^\nu}{\rho}$.[22]

Specific values of D^e defined by Eq. (A.41) will be discussed after deriving the expression of the electron-energy transport in the following.

A.4.2 Electron-Energy Transport

Analogous to the electron diffusion, the electron-energy transport is defined herein as that relative to the average motion of the gas mixture. Hence we seek the quantity

$$q_\alpha^e = \int f_e(\tfrac{1}{2} m^e C_e^2) C_\alpha^e d\vec{C}^e \quad \text{(A.46)}$$

for the small but nonvanishing values of the Knudsen number ϵ.

Using Eqs. (A.25), (A.30), and (A.36), Eq. (A.46) becomes

$$\begin{aligned} q_\alpha^e = \tfrac{1}{2} m^e \Bigg[&-\left(\frac{1}{p^e}\frac{\partial p^e}{\partial x_\beta} + \frac{eE_\beta}{kT^e} \right) \int f_{e0} \frac{C_e^2 C_\alpha^e C_\beta^e}{C_e(n^i Q^{ei} + n^a Q^{ea})} d\vec{C}^e \\ &- \frac{1}{T^e}\frac{\partial T^e}{\partial x_\beta} \int f_{e0} \frac{C_e^2 C_\alpha^e C_\beta^e [(m^e C_e^2)/(2kT^e) - 5/2]}{C_e(n^i Q^{ei} + n^a Q^{ea})} d\vec{C}^e \\ &+ \frac{m^e}{kT^e}\frac{\partial u_\gamma}{\partial x_\beta} \int f_{e0} \frac{C_e^2 (C_\gamma^e C_\beta^e - \tfrac{1}{3} C_e^2 \delta_{\gamma\beta}) C_\alpha^e}{C_e(n^i S^i + n^a S^a)} d\vec{C}^e \Bigg] \end{aligned} \quad \text{(A.47)}$$

The last term within the bracket of the above equation is zero for the same reason that the last term of Eq. (A.39) was zero.

After some manipulation Eq. (A.47) is put into the form

$$q_\alpha^e = -\frac{5}{2} n^e k T^e D^e \left(\frac{1}{p^e}\frac{\partial p^e}{\partial x_\alpha} + \frac{eE_\alpha}{kT^e} \right) - K_h^e \frac{\partial T^e}{\partial x_\alpha} \quad \text{(A.48)}$$

where the electron thermal conductivity K_h^e is defined as

$$K_h^e = \frac{m^e}{6T^e} \int f_{e0} \frac{C_e^3 [(m^e C_e^2)/(2kT^e) - 5/2]}{(n^i Q^{ei} + n^a Q^{ea})} d\vec{C}^e \quad \text{(A.49)}$$

Note that the integral defining K_h^e, which comes from the second integral of

[22] Because we have used C to denote molecular velocity, N is employed to denote mass fraction.

Eq. (A.47), is not the same as the integral in Eq. (A.39) defining the thermal diffusion. Rather, the thermal diffusion integral of Eq. (A.39) appears in the course of the integration of the first integral of Eq. (A.47), and its contribution is neglected in obtaining Eq. (A.48). Therefore, Eq. (A.48) is valid when the thermal diffusion of electrons is negligible. That portion of the pressure diffusion that is proportional to the total pressure gradient is also neglected in deriving Eq. (A.48), consistent with the derivation of Eq. (A.40).

Equation (A.48) shows that the electron-energy transport is equal to the sum of that caused by the diffusional transport of the electron enthalpy, $\frac{5}{2} kT^e$, and that caused by the thermal conduction.

The values of K_h^e will be discussed in the following section.

A.4.3 Coefficients of Diffusion and Thermal Conductivity

The electron diffusion coefficient and thermal conductivity are given by Eqs. (A.41) and (A.49), respectively. The integrals of these equations can be evaluated essentially in closed form for the inverse-power intermolecular force laws. For other force laws, the integration is rather complicated.

The Coulomb cross section governing the collisions between the charged particles Q^{ei} is about one hundred times greater than the electron-neutral collision cross section Q^{ea} for the usual temperatures of interest of the electrostatic probe operations.[23] The electron-neutral cross section is of the same order of magnitude as that between the neutral gas particles. With this information, let us consider the quantity, $(n^i Q^{ei} + n^a Q^{ea})$, appearing in Eqs. (A.41) and (A.49). If a plasma is so weakly ionized that

$$\frac{n^i Q^{ei}}{n^a Q^{ea}} \leq 0.1 \tag{A.50}$$

then we see that $n^i Q^{ei}$ is negligible as compared to $n^a Q^{ea}$ in the quantity $(n^i Q^{ei} + n^a Q^{ea})$. Under such circumstances, the plasma behavior is governed by the collisions between the ionized and the neutral species, and Coulomb collisions between the ionized species can be ignored. The plasma whose degree of ionization satisfies the criterion of Eq. (A.50) is referred in probe work as a *weakly ionized* plasma. The degree of ionization, therefore, of a "weakly ionized" plasma is usually much less than 1 percent.

In contrast, a "strongly ionized" plasma is a plasma for which degree of ionization is sufficiently high such that

$$\frac{n^i Q^{ei}}{n^a Q^{ea}} \gg 1 \tag{A.51}$$

The Coulomb collisions determine the plasma properties for such a plasma.

[23] In certain special cases Q^{ea} is much smaller than this, e.g., in argon in certain temperature ranges owing to the Ramsauer effect.

We have seen that for plasmas that are weakly ionized only $n^a Q^{ea}$ is required in evaluating the diffusion coefficient and the thermal conductivity. However, for strongly ionized plasmas only $n^i Q^{ei}$ is required. The integrals of Eqs. (A.41) and (A.49) can be readily integrated in closed form for the weakly or the strongly ionized plasmas when an inverse intermolecular power law is employed. Sutton and Sherman (1965) give the following expressions:

$$D^e = \frac{2}{3A_1(s)} \frac{kT^e}{\pi^{3/2} m^e n^\nu} \left(\frac{m^e}{2kT^e}\right)^{1/2} \left(\frac{2kT^e}{|F_0^{e\nu}|}\right)^{2/s-1} \Gamma\left(\frac{2s}{s-1}\right) \tag{A.52}$$

and

$$K_h^e = \frac{k}{3\pi^{3/2} A_1(s)} \frac{n^e}{n^\nu} \left(\frac{m^e}{|F_0^{e\nu}|}\right)^{2/(s-1)} \left(\frac{2kT^e}{m^e}\right)^{(s+3)/2(s-1)}$$
$$\left[\Gamma\left(\frac{4s-2}{s-1}\right) - \frac{5}{2}\Gamma\left(\frac{3s-1}{s-1}\right)\right] \tag{A.53}$$

where $\nu = a$ for weakly ionized plasmas, and $\nu = i$ for strongly ionized plasmas. The symbol $\Gamma(\ )$ represents the gamma function.

$F_0^{e\nu}$ and s are the constant and the exponent, respectively, appearing in the inverse-power intermolecular force law. $A_1(s)$ represents the value of an integral that appears during the course of evaluation of the integrals of Eqs. (A.41) and (A.49). Values of $A_1(s)$ are given by Chapman and Cowling (1960) and are also cited by Sutton and Sherman (1965). The constant $F_0^{e\nu}$ of the intermolecular force law is usually determined by making the calculated value with experimental data at some reference condition.

We see from Eqs. (A.52) and (A.53) that both D^e and K_h^e vary linearly with T^e when $s = 5$, and vary as $(T^e)^{1/2}$ when $s = \infty$. The latter two values of s correspond to the so-called Maxwell and hard-sphere molecules, respectively. These particular intermolecular force laws are appropriate for the electron-neutral collisions, and therefore the expressions for D^e and K_h^e for Maxwell and hard-sphere molecules apply to weakly ionized plasmas.

For strongly ionized plasmas, the Coulomb collisions are dominant for which the appropriate exponent for the intermolecular force law is $s = 2$. For this case, $A_1(2)$ is a function of T_e. However, $A_1(2)$ is a weak function of T_e, so that D^e and K_h^e vary essentially with $(T^e)^{5/2}$. The reader is referred to Spitzer (1962) for a discussion of the strongly ionized case.

A.4.4 Transport Rates of Ions and Ion Thermal Energy

We have seen earlier in Eq. (A.21) that the zeroth-order distribution function for the ions f_{i0} is the Maxwellian distribution function about the temperature T, which is equal to the neutral gas temperature. The solution for f_{i1} can be obtained in the same manner as that for f_{e1} and results in an equation for f_{i1} that is similar to Eq. (A.36). Because all the temperature terms that appear in Eq. (A.36) involve the electron temperature T^e and arise from the fact that T^e is contained in f_{e0}, one can readily see that all temperatures appearing in f_{i1} will be the neutral gas temperature T. Therefore the ion temperature is equal to the neutral gas temperature up through first order in ϵ. This means that the ions are always in thermal equilibrium with the neutral molecules in plasma flow regimes where the continuum analysis is valid.

The transport of ions and of ion energy are obtained by the use of f_{i1} in the same manner as that in which we obtained the transport of electrons and of electron energy by the use of f_{e1}. The results are

$$m^i n^i \langle C_\alpha^i \rangle = -m^i n^i D^i \left(\frac{1}{p^i} \frac{\partial p^i}{\partial x_\alpha} - \frac{eZ}{kT} E_\alpha - \frac{m^i}{\rho kT} \frac{\partial P}{\partial x_\alpha} \right) \tag{A.54}$$

and

$$q_\alpha^i = -\left[K_h^i \left(\frac{\partial T}{\partial x_\alpha} \right) + \left(\frac{5}{2} n^i k T^i \right) \langle C_\alpha^i \rangle \right] \tag{A.55}$$

where thermal diffusion has again been neglected, analogous to Eqs. (A.40) and (A.48). Note that the term $(m^i/\rho kT\ \partial p/\partial x_\alpha)$ is not neglected in contrast to the corresponding term of Eq. (A.39), since $m^i \gg m^e$.

As was done for electron diffusion, the ion mobility K^i can be defined by the Einstein relationship as

$$K^i = D^i \frac{e}{kT} \tag{A.56}$$

Equation (A.54) can be put into the more familiar form of Eq. (A.45) when the total pressure is constant in the direction of diffusion and $T = T^e$; as

$$m^i n^i \langle C_\alpha^i \rangle = -\rho \left(D^i \frac{\partial N^i}{\partial x_\alpha} - Z K^i E_\alpha N^i \right) \tag{A.57}$$

where N^i is the mass fraction of ions.

The expressions for the diffusion coefficient and the thermal conductivity for ions D^i and K_h^i are basically the same as those for electrons given by Eqs. (A.52) and (A.53), except that m^e and T^e are changed to m^i and T^i, respectively. Also, the numerical constants of the expressions for ions are somewhat different from those for electrons depending on the values of s. The species ν in the expressions for D^i and K_h^i is the neutral gas when the plasma is weakly ionized whereas it is the ions when the plasma is strongly ionized.

The constant $F_0^{e\nu}$ appearing in the inverse-power intermolecular force law is a function of the particle masses of the colliding pair as $m^e m^\nu$. Hence Eq. (A.52) shows that for the Maxwell molecules ($s = 5$)

$$\frac{D^i}{D^e} \propto \left(\frac{T}{T^e} \right) \left(\frac{m^e}{m^i} \right) \tag{A.58}$$

whereas for the hard-sphere model ($s = \infty$),

$$\frac{D^i}{D^e} \propto \left(\frac{T}{T^e}\frac{m^e}{m^i}\right)^{1/2} \tag{A.59}$$

The numerical constant appearing in the expression for D^i is $\sqrt{2}$ times that for D^e. Hence the constants of proportionality of Eqs. (A.58) and (A.59) are both $\sqrt{2}$.

Similarly, one can show from Eq. (A.53) that, for $s = 5$

$$\frac{K_h^i}{K_h^e} \propto \frac{n^i}{n^e}\frac{T}{T^e}\frac{m^e}{m^i} \tag{A.60}$$

whereas for $s = \infty$

$$\frac{K_h^i}{K_h^e} \propto \frac{n^i}{n^e}\left(\frac{T}{T^e}\frac{m^e}{m^i}\right)^{1/2} \tag{A.61}$$

Both $s = 5$ and $s = \infty$ apply to collisions of the ionized species with the neutral species. Hence Eqs. (A.58)–(A.61) are for weakly ionized plasmas. Similar relationships can also be deduced for strongly ionized plasmas from Eqs. (A.52) and (A.53) by setting $s = 2$.

A.5 Moment Equations of the Boltzmann Equations

In this section we shall first derive the generalized moment equations corresponding to the Boltzmann equations, Eqs. (A.1)–(A.3). We shall deduce, from these generalized moment equations, a set of lowest-order moment equations, which become closed for the continuum regime when we make use of the expressions for the various transport rates derived previously. These closed lowest order moment equations constitute the governing equations of continuum plasma flows that were analyzed in Chapter III.

A.5.1 The Generalized Moment Equation

We let θ^ν be a generalized function of $\vec{\xi}^\nu$. We multiply Eqs. (A.1)–(A.3) by θ^e, θ^i, θ^a, respectively. Then, as we integrate these equations with respect to $\vec{\xi}^e$, $\vec{\xi}^i$, and $\vec{\xi}^a$, respectively, three generalized moment equations result, which can be represented by the following equation.

$$\frac{\partial}{\partial t}\int f_\nu \theta^\nu d\vec{\xi}^\nu + \frac{\partial}{\partial x_\alpha}\int f_\nu \xi_\alpha^\nu \theta^\nu d\vec{\xi}^\nu - \Delta^\nu \frac{e}{m^\nu} E_\alpha \int \theta^\nu \frac{\partial f_\nu}{\partial \xi_\alpha^\nu} d\vec{\xi}^\nu$$
$$= \sum_{\bar{\nu}=e,i,a} \iiiint (\theta^{\nu\prime} - \theta^\nu) f_\nu f_{\bar{\nu}} g b\, db\, d\varepsilon\, d\vec{\xi}^{\bar{\nu}} d\vec{\xi}^\nu \tag{A.62}$$

where $\Delta^e = 1$, $\Delta^i = -1$, and $\Delta^a = 0$. Also, α, β, and γ are exclusively reserved for the tensor indices.

Equation (A.62) and the particular moment equations to be derived subsequently from this equation are valid for all plasma-flow regimes. We shall, however, examine the particular forms of these equations in the continuum limit.

A.5.2 Continuum Conservation Equations for Electrons and Ions

The conservation equations of electrons, ions, and of the global mixture will be derived for the continuum regime.

We let $\theta^e = m^e$, $\theta^i = m^i$, and $\theta^a = m^a$, successively in Eq. (A.62) and obtain

$$\frac{\partial \rho N^e}{\partial t} + \frac{\partial}{\partial x_\beta}[\rho N^e(u_\beta + \langle C_\beta^e \rangle)] = \omega^e \tag{A.63}$$

$$\frac{\partial \rho N^i}{\partial t} + \frac{\partial}{\partial x_\beta}[\rho N^i(u_\beta + \langle C_\beta^i \rangle]) = \omega^i \tag{A.64}$$

$$\frac{\partial \rho N^a}{\partial t} + \frac{\partial}{\partial x_\beta}[\rho N^a(u_\beta + \langle C_\beta^a \rangle)] = \omega^a \tag{A.65}$$

where ω^ν denotes the mass rate of production per unit volume of the species ν by chemical reaction. We note that

$$\sum_\nu N^\nu = 1, \quad \sum_\nu \omega^\nu = 0, \quad \text{and} \quad \sum_\nu N^\nu \langle C_\alpha^\nu \rangle = 0 \tag{A.66}$$

we add the three equations, Eqs. (A.63)–(A.65), and obtain the global mass conservation equation

$$\frac{\partial \rho}{\partial t} + \frac{\partial \rho u_\beta}{\partial x_\beta} = 0 \tag{A.67}$$

With the use of the above equations, Eqs. (A.63) and (A.64) become

$$\rho \frac{\partial N^e}{\partial t} + \rho u_\beta \frac{\partial N^e}{\partial x_\beta} + \frac{\partial}{\partial x_\beta}(\rho N^e \langle C_\beta^e \rangle) = \omega^e \tag{A.68}$$

$$\rho \frac{\partial N^i}{\partial t} + \rho u_\beta \frac{\partial N^i}{\partial x_\beta} + \frac{\partial}{\partial x_\beta}(\rho N^i \langle C_\beta^i \rangle) = \omega^i \tag{A.69}$$

The closure of the above equations is now accomplished for the continuum regime by substituting the continuum expressions, Eqs. (A.40) and (A.54), for $\langle C_\alpha^e \rangle$ and $\langle C_\alpha^i \rangle$. Noticing that $\rho N^\nu = m^\nu n^\nu$, Eqs. (A.68) and (A.69) for the continuum plasma regime finally reduce to

$$\rho \frac{\partial N^e}{\partial t} + \rho u_\beta \frac{\partial N^e}{\partial x_\beta} - \frac{\partial}{\partial x_\beta}\left[\rho D^e\left(\frac{N^e}{p^e}\frac{\partial p^e}{\partial x_\beta} + \frac{e}{kT^e}E_\beta N^e\right)\right] = \omega^e \tag{A.70}$$

and

$$\rho \frac{\partial N^i}{\partial t} + \rho u_\beta \frac{\partial N^i}{\partial x_\beta} - \frac{\partial}{\partial x_\beta}\left[\rho D^i\left(\frac{N^i}{p^i}\frac{\partial p^i}{\partial x_\beta} - \frac{Ze}{kT}E_\beta N^i - \frac{N^i m^i}{\rho kT}\frac{\partial P}{\partial x_\beta}\right)\right] = \omega^i \quad \text{(A.71)}$$

where thermal diffusion effects have been neglected as per the discussion following Eq. (A.39).

A.5.3 Continuum Electron-Energy Equation

To obtain the continuum electron-energy equation we form the moment of the electron energy by substituting in Eq. (A.62)

$$\theta^e = \tfrac{1}{2} m^e \xi^e_\beta \xi^e_\beta \quad \text{(A.72)}$$

The following equation then results after some manipulation.

$$\begin{aligned}
&\frac{\partial}{\partial t}\rho^e(h^e + u_\beta\langle C^e_\beta\rangle) - \frac{\partial p^e}{\partial t} + \frac{\partial}{\partial x_\beta}(\rho^e h^e u_\beta) + \frac{\partial}{\partial x_\beta}(q^e_\beta) - \rho^e\left(\frac{e}{m^e}\right)\frac{\partial \varphi}{\partial x_\beta}\langle C^e_\beta\rangle \\
&\quad - \rho^e\left(\frac{e}{m^e}\right)\frac{\partial \varphi}{\partial x_\beta} u_\beta + \frac{\partial}{\partial x_\beta}\rho^e[\tfrac{1}{2}(u_\gamma u_\gamma)\langle C^e_\beta\rangle + u_\beta(u_\gamma\langle C^e_\gamma\rangle)] \\
&\quad + \frac{\partial}{\partial x_\beta}\rho^e u_\gamma(\langle C^e_\gamma C^e_\beta\rangle - \tfrac{1}{3}\langle C^2_e\rangle\delta_{\gamma\beta}) \\
&\quad = \sum_{\bar{\nu}=i,a} \tfrac{1}{2} m^e \int\int\int\int [(\xi^2_e)' - \xi^2_e] f_e f_{\bar{\nu}} g b\, db\, d\bar{\epsilon}\, d\bar{\xi}^{\bar{\nu}} d\bar{\xi}^e
\end{aligned} \quad \text{(A.73)}$$

In the above equation the stagnation enthalpy of the electrons h^e and the electric potential φ are defined as

$$h^e = \frac{5}{2}\frac{\langle C^e_\gamma C^e_\gamma\rangle}{3} + \tfrac{1}{2}u_\gamma u_\gamma = \frac{5}{2}\frac{kT^e}{m^e} + \frac{u^2}{2} \quad \text{(A.74)}$$

and

$$E_\alpha = -\frac{\partial \varphi}{\partial x_\alpha}$$

Because of the extremely small electronic mass, the last two terms on the left-hand side of Eq. (A.73) usually can be neglected.

Closure of Eq. (A.73) is now accomplished for continuum plasma flows by substituting Eq. (A.48), the expression for the continuum electron-energy transport q^e_β, which was derived earlier.

Before we transform Eq. (A.73) into a more convenient form, let us consider the collision term on the right-hand side of that equation, which, for the purpose of the present discussion we denote by I_c. To be consistent, we must employ the expressions we developed earlier for f_e and f_ν which are accurate through first order in ϵ in the evaluation of the collision integral I_c.

For the moment of Eq. (A.62) under consideration, $\theta^e = \frac{1}{2}m^e C_e^2$, however, this integral does not vanish for the leading term of $f_e f_\nu$, which is $f_{e0} f_{\nu 0}$. Because the next-order term of $f_e f_\nu$, which is $(f_{e0} f_{\nu 1} + f_{e1} f_{\nu 0})$, is order ϵ as compared to $f_{e0} f_{\nu 0}$, the collision integral should be evaluated with the use of the leading term of $f_e f_\nu$, namely the Maxwellian functions $f_{e0} f_{\nu 0}$.

The integral I_c on the right-hand side of Eq. (A.73) with $f_e f_\nu$ set equal to $f_{e0} f_{\nu 0}$ can be evaluated in closed form if a simple intermolecular force law, such as the inverse fifth-power law, is assumed (Chung, 1969). For more general intermolecular force laws, the integration is rather difficult.

In any case, the value of the integral I_c is negligibly small for any of the elastic collision processes. It is appreciable only for inelastic collisions. For inelastic collisions, however, the value of the integral depends among other things on the particular internal degrees of freedom of the heavy gas molecules with which the electrons exchange energy. Therefore, it cannot be evaluated in a straightforward manner. It can, however, in general be represented approximately by the expression

$$I_c = 3\rho^e k \sum_{\bar{\nu}=i,a} \frac{1}{m^{\bar{\nu}}} a^{\bar{\nu}} \tau^{e\bar{\nu}} (T^e - T) \tag{A.75}$$

where $\tau^{e\bar{\nu}}$ is the approximate average collision frequency for $e - \nu$ collisions. It is given, for the situations most commonly encountered in probe problems wherein the electron-diffusion velocity is much smaller than the average thermal velocity of the electrons, by

$$\tau^{e\bar{\nu}} = \frac{n^{\bar{\nu}}}{3n^e k T^e} \int f_{e0} Q^{e\bar{\nu}} C_e^3 d\vec{C}^e \tag{A.76}$$

The factor $a^{\bar{\nu}}$ is unity for elastic collisions. For inelastic collisions, it is much greater than unity and depends on the particular internal degrees of freedom of the heavy gas particles which exchange energy with the electrons.

Before we leave the subject of the collision term I_c on the right-hand side of Eq. (A.73), let us dwell for a moment on the meaning of the term. This term represents the collisional energy exchange between the electrons and the other gas particles of the plasma. This collisional energy exchange is equal to zero according to the Lorentz approximation, which we have adhered to in the first-order continuum solution of the Boltzmann equations. Since the closure of Eq. (A.73) is accomplished by using the expression for q_β^e which was derived from this first-order solution of Boltzmann's equations, the inclusion of the energy-exchange term at this point is not quite consistent. It is, however, customary to discard the Lorentz approximation and include this term in the final formulation of the continuum energy equations.

Now, Eq. (A.73) is put into the following form with the use of Eqs. (A.48) and (A.68).

$$\frac{\partial}{\partial t}(\rho^e u_\beta \langle C_\beta^e \rangle) - \frac{\partial p^e}{\partial t} + \frac{e}{m^e}\rho^e\frac{\partial \varphi}{\partial t} + \rho^e\frac{\partial H^e}{\partial t} + \rho^e(\langle C_\beta^e \rangle + u_\beta)\frac{\partial H^e}{\partial x_\beta}$$

$$- \frac{\partial}{\partial x_\beta}\left(K_h^e \frac{\partial T^e}{\partial x_\beta}\right) = -\omega^e h^e + 3\rho^e k \sum_{\bar{\nu}=i,a}\left(\frac{1}{m^{\bar{\nu}}}\right) a^{\bar{\nu}} \tau^{e\bar{\nu}}(T - T^e) \qquad \text{(A.77)}$$

where $\langle C_\beta^e \rangle$ is given by Eq. (A.40), and the total enthalpy of electrons, H^e, is defined by

$$H^e = h^e - \frac{e\varphi}{m^e} = \tfrac{1}{2}u^2 + \frac{5}{2}\frac{kT^e}{m^e} - \frac{e\varphi}{m^e} \qquad \text{(A.78)}$$

where $u^2 = u_\gamma u_\gamma$.

For steady state with frozen energy exchange and frozen chemical reactions, the electron-energy equation becomes

$$\rho^e \langle \xi_\beta^e \rangle \frac{\partial H^e}{\partial x_\beta} = \frac{\partial}{\partial x_\beta}\left(K_h^e \frac{\partial T^e}{\partial x_\beta}\right) \qquad \text{(A.79)}$$

where $\langle \xi_\beta^e \rangle$ is the absolute average velocity of the electrons. This equation shows that convection of the total enthalpy defined by Eq. (A.78) is balanced by the electron-energy conduction. Note that T^e is defined in terms of $\langle C_\gamma^e C_\gamma^e \rangle$ where C_γ^e is the velocity relative to the mass averaged velocity u_γ and not $\langle \xi_\gamma^e \rangle$. Therefore $\frac{5}{2}\frac{kT^e}{m^e} + \frac{1}{2}u^2$ represents all of the electron energy, except that due to φ.

A.5.4 Continuum Heavy-Gas Energy Equation

Since, as it was shown earlier, the ion temperature is equal to the neutral gas temperature in the continuum regime, we need not consider the ion-energy equation separately. Rather, we shall derive the energy equation of the gas mixture excluding electrons.

We substitute successively in Eq. (A.62)

$$\theta^i = \tfrac{1}{2}m^i \xi_\alpha^i \xi_\alpha^i$$

and

$$\theta^a = \tfrac{1}{2}m^a \xi_\alpha^a \xi_\alpha^a \qquad \text{(A.80)}$$

Two equations similar to Eq. (A.73) result, which can be represented by the following equation for $\nu = i$ and a.

$$\frac{\partial}{\partial t}\rho^\nu(h^\nu + u_\alpha\langle C_\alpha^\nu \rangle) + \frac{\partial p^\nu}{\partial t} + \frac{\partial}{\partial x_\beta}(\rho^\nu h^\nu u_\beta) + \frac{\partial}{\partial x_\beta}(q_\beta^\nu) + \Delta^\nu \rho^\nu \frac{e}{m^\nu}\frac{\partial \varphi}{\partial x_\beta}(\langle C_\beta^\nu \rangle + u_\beta)$$

$$+ \frac{\partial}{\partial x_\beta}\rho^\nu[\tfrac{1}{2}u^2\langle C_\beta^\nu \rangle + u_\beta(u_\alpha\langle C_\alpha^\nu \rangle)] + \frac{\partial}{\partial x_\beta}\rho^\nu u_\alpha(\langle C_\alpha^\nu C_\beta^\nu \rangle - \tfrac{1}{3}\langle C_\nu^2 \rangle \delta_{\alpha\beta})$$

$$= \tfrac{1}{2}m^\nu \int\!\!\int\!\!\int\!\!\int [(\xi_\nu^2)' - \xi_\nu^2] f_\nu f_e g b\, db\, d\bar{\varepsilon}\, d\bar{\xi}^\nu d\bar{\xi}^e \qquad \text{(A.81)}$$

where the definition analogous to Eq. (A.74) applies to h^ν.

Since ρ^e is negligibly small as compared to ρ^i and ρ^a, we may write for all practical purposes

$$\rho = \sum_{\nu=e,i,a} \rho^\nu \approx \rho^i + \rho^a$$
$$\sum_{\nu=e,i,a} \rho^\nu \langle C_\alpha^\nu \rangle \approx \rho^i \langle C_\alpha^i \rangle + \rho^a \langle C_\alpha^a \rangle = 0 \tag{A.82}$$

We now add the two equations represented by Eq. (A.81), and make use of Eqs. (A.82), and close the resulting equation by the use of the continuum expressions for q_α^ν and the shear stresses. We then obtain

$$\frac{\partial p^{ia}}{\partial t} + \rho \frac{\partial h}{\partial t} + \rho u_\beta \frac{\partial h}{\partial x_\beta} + \frac{\partial}{\partial x_\beta}(q_\beta^{ia}) + \frac{\partial}{\partial x_\beta}(u_\alpha \zeta_{\alpha\beta}^{ia})$$
$$= -\frac{eZ}{m^i} \rho^i E_\beta \xi_\beta^i + 3\rho^e k \sum_{\bar{\nu}=i,a} \frac{1}{m^{\bar{\nu}}} a^\nu \tau^{e\bar{\nu}} (T^e - T) \tag{A.83}$$

where

$$p^{ia} = p^i + p^a$$
$$h = h^i = h^a$$
$$q^{ia} = q_\beta^i + q_\beta^a = (K_h^i + K_h^a) \frac{\partial T}{\partial x_\beta}$$
$$\zeta_{\alpha\beta}^{ia} = \sum_{\nu=i,a} (\rho^\nu \langle C_\alpha^\nu C_\beta^\nu \rangle - p^\nu \delta_{\alpha\beta}) \tag{A.84}$$

The shear stress $\zeta_{\alpha\beta}^{ia}$ is proportional to the usual average velocity gradients, and the viscosity depends on the collision properties of the ions and neutrals.

The basic form of Eq. (A.83) is the same as that for a mixture of neutral gases except for the terms of the right-hand side. The first term on the right is the joule heating term, and the second term represents the collisional energy exchange with the electrons.

For weakly ionized plasmas where $\rho^i \ll \rho^a$, we have $K_h^i \ll K_h^a$, and also the viscosity becomes a function only of the collision properties of the neutrals. Furthermore, the right-hand side of Eq. (A.83) becomes negligible for $\rho^i \ll \rho^a$. The heavy-gas energy equation, Eq. (A.83), then becomes identical to the usual continuum energy equation for the neutral species, and it becomes decoupled from the electrical aspect of the plasma.

A.5.5 Plasma Momentum Equation

By substituting $\theta^e = m^e \xi_\alpha^e$, $\theta^i = m^i \xi_\alpha^i$, and $\theta^a = m^a \xi_\alpha^a$, successively into Eq. (A.62), we generate a set of three species momentum equations. Each of these equations has nonvanishing collision terms. As we add these three species equations together, however, the sum of the collision terms vanishes since the total momentum of the species comprising the plasma is a collisional invariant. In the continuum regime, the resulting global momentum equation is the regular Navier–Stokes equation wherein the viscosity is evaluated from the collision properties between all the species comprising the plasma. The momentum equation is, therefore, the same as that of a mixture of electrically neutral gases.

A.5.6 Poisson Equation and Closing Remarks on Continuum Regime

The set of the governing equations becomes complete for the continuum flow regime as we add the Poisson equation, Eq. (A.4), to the six conservation equations we have discussed. The Poisson equation is written here as

$$\frac{\partial E_\beta}{\partial x_\beta} = \frac{e}{\sigma}(Zn^i - n^a) \tag{A.85}$$

The six conservation equations we discussed for the continuum regime are the global continuity equation, Eq. (A.67), the vector Navier–Stokes equation, the electron- and ion-conservation equations, Eqs. (A.70) and (A.71), the electron-energy equation, Eq. (A.77), and the heavy gas energy equation, Eq. (A.83). These conservation equations represent the first part of the solution of the Boltzmann equations for small values of Knudsen number, that is, for the continuum regime. In order to complete the solution, the probe geometry, the free-stream plasma properties, and the probe surface conditions must be specified. Solutions of these continuum conservation equations with the appropriate boundary conditions specifying the probe geometry, etc., have been discussed in Chapter III.

A.6 Noncontinuum Regimes (Collisionless and Transitional)

Obviously, for the regimes where the Knudsen number is of order one or greater, the expansion scheme of Eq. (A.16) fails, and one must resort to other methods of solution of the Boltzmann equations, Eqs. (A.1)–(A.3).

In the completely collisionless regimes, cases 1 and 2 under heading A of the list given in Chapter I, exact solutions of the Boltzmann–Vlasov equations can be obtained.

For all other regimes of Knudsen number of order one or greater, the full Boltzmann equations, Eqs. (A.1)–(A.3), must be solved along with the Poisson equation, at least for certain portions of the flow field.

Solution of the full Boltzmann equations is a very complicated mathematical problem. Only a few approximate solutions of Eqs. (A.1)–(A.4) have been obtained thus far. Most of these solutions employ certain modifications of the moment method due to Mott-Smith (1951), Liu and Lees (1961), and/or approximations to the collision integral such as that introduced by Krook

(1955). A discussion of such solutions would take us too far afield, and we must refer the interested reader to such texts as Mitchner and Kruger (1973) and Shkarofsky *et al.* (1966), as well as the other references already cited in this Appendix.

References

Balescu, R. (1960), *Phys. Fluids*, **3**, 52.

Burke, A. F. (1967), CAL Inc. Rep. AN-2101-Y-1, May 1967.

Chapman, S. and Cowling, T. G. (1960), *The Mathematical Theory of Non-Uniform Gases*, Chaps. 7, 8, and 18, Cambridge Univ. Press, England.

Chung, P. M. (1969), *Phys. Fluids*, **12**, 1623.

Hirschfelder, J. O., Curtiss, C. F., and Bird, R. B. (1954), *The Molecular Theory of Gases and Liquids*, Chapter 7, Wiley, New York.

Jeans, J. H. (1925), *The Dynamical Theory of Gases*, 4th ed., Chaps. 8 and 9, Dover, New York.

Krook, M. (1955), *Astrophys. J.*, **122**, 488.

Lenard, A. (1960), *Ann. Phys.*, **10** (3), 390.

Liu, C. Y. and Lees, L. (1961), *Rarefied Gas Dynamics* (L. Talbot, ed.), pp. 391–428, Academic Press, New York.

Mitchner, M. and Kruger, C. H., Jr. (1973), *Partially Ionized Gases*, Wiley, New York.

Mott-Smith, H. M. (1951), *Phys. Rev.*, **82**, 885.

Shkarofsky, I. P. T., Johnston, T. W., and Bachynski, M. P. (1966), *The Particle Kinetics of Plasmas*, Addison-Wesley, Reading, Massachusetts.

Spitzer, L., Jr. (1962), *Physics of Fully Ionized Gases*, Wiley-Interscience, New York.

Sutton, G. W. and Sherman, A. (1965), *Engineering Magnetohydrodynamics*, Chaps. 4 and 5, McGraw-Hill, New York.

Index

MIX
Papier aus verantwortungsvollen Quellen
Paper from responsible sources
FSC® C105338

If you have any concerns about our products,
you can contact us on
ProductSafety@springernature.com

In case Publisher is established outside the EU,
the EU authorized representative is:
Springer Nature Customer Service Center GmbH
Europaplatz 3, 69115 Heidelberg, Germany

Printed by Libri Plureos GmbH
in Hamburg, Germany